A WING
AND
A PRAYER

The Race to Save Our Vanishing Birds

ANDERS AND BEVERLY GYLLENHAAL

SIMON & SCHUSTER

New York   London   Toronto   Sydney   New Delhi

Simon & Schuster
1230 Avenue of the Americas
New York, NY 10020

First Simon & Schuster hardcover edition April 2023

SIMON & SCHUSTER and colophon are registered trademarks of Simon & Schuster, Inc.

For information about special discounts for bulk purchases, please contact
Simon & Schuster Special Sales at 1-866-506-1949
or business@simonandschuster.com.

The Simon & Schuster Speakers Bureau can bring authors to your live event.For more
information or to book an event, contact the
Simon & Schuster Speakers Bureau at 1-866-248-3049
or visit our website at www.simonspeakers.com.

Interior design by Ruth Lee-Mui

Manufactured in the United States of America

1   3   5   7   9   10   8   6   4   2

Library of Congress Control No. 1982184558

ISBN 978-1-9821-8455-1
ISBN 978-1-9821-8457-5 (ebook)

For Sam and Grey, with love

Contents

Introduction: What the Birds Are Telling Us   1

1. On the Edge of Extinction   11
2. Vanishing by the Billions   33
3. Era of Discovery   51
4. Following the Birds   63
5. Listening to the Birds   85
6. Canary in the Coal Mine   105
7. World Travelers   123
8. When All Else Fails   145
9. Coexisting with the Birds   171
10. Case Studies in Getting It Done   191

Conclusion: Making the Case for Birds   217

Afterword: How You Can Help   229
Acknowledgments   243
Notes   247
Selected Bibliography   297
Photo Credits   299
Index   301

Bluebirds, among the most beloved of common species, could be the symbol of the modern bird crisis. Their populations plummeted through much of the last century. Today, the health of bluebirds, like this Eastern Bluebird, rests on the willingness of people to help them along. The North American Bluebird Society and local chapters are leading efforts to build nest boxes to provide homes for the birds throughout the continent.

WHAT THE BIRDS ARE TELLING US

"There he is," someone shouts, and sure enough the tiny sharp beak of Florida Grasshopper Sparrow No. 2050176 pokes out from behind a clump of wiregrass. Ever so slowly, the bird steps forward until he reaches the ledge of the giant mobile cage where he spent the past day getting ready for the mission ahead. Hatched in captivity and raised for this very moment, the bird woke up to something he's never seen before: The enclosure's front metal wall is gone and an ocean of Central Florida grassland is spread out before him.

A handful of researchers watch crouched in the surrounding field, barely breathing as they wait to see if twenty years of research and experimentation will pay off. A full minute, then another tick by while the small brown bird, its spindly pink legs ringed with ID bands, just stands there. At last the sparrow makes his move. He half flies, half dives off the ledge into the wild in what everyone hopes will be the first step in rebuilding the continent's most endangered bird.

At the other end of the country, a foot-and-a-half-tall California Spotted Owl is planted on her nest on a jagged, broken branch in the most remote corner of the Sierra Nevada mountain range. This is rugged country, filled with six varieties of towering pines and canyons carved out by a half dozen rivers, punctuated by giant boulders left scattered by glaciers four million years ago. The closest roads, more like dirt paths, get almost no traffic. The only sounds many days are chattering jays and juncos and the wind whooshing by at 6,000 feet above sea level.

There aren't any people around, but the owl isn't exactly alone. Her every move is captured three different ways. When the owl launches into her loud, barklike call, the hoots are picked up by high-tech recorders strapped to nearby trees. When she shoots down to snag a flying squirrel, the hunt is captured by a motion-activated camera. When the owl leaves the nest to patrol her territory, a tiny transmitter attached to her tail feathers comes along. Every hoot, whistle, and call is tracked and analyzed in the world's largest project using sound to study wildlife. In a few months, when researchers collect what amounts to a million hours of recordings from throughout the entire range, they hope to know what it will take to save this owl.

These two projects serve as bookends of the extraordinary work under way to protect birds across the hemisphere. The Grasshopper Sparrow is among the least-known and latest to brush up against extinction, while the Spotted Owl is among the best-known and part of the most storied chapter in the history of birds in the United States. Between them lies an array of rescue missions—some well financed, others threadbare, some succeeding, others losing ground—that will help determine the future of North America's birdscape.

Birds are the most visible branch of wildlife, found in every corner of the globe and all too easy to take for granted. We certainly did during our first decade as birders, content to see them as gifts of nature here for our delight. But a series of advances in the science and technology of bird research recently led to a startling discovery. In the past fifty years nearly a third of the bird population in North America has withered away, up against the loss of habitat, shifting climate, and growing hazards of an urban world.

That translates to three billion birds of all sizes and shapes, in losses stretching from coast to coast, from the Arctic to Antarctica, through forests and grasslands, ranches and farms. As one veteran biologist, John Doresky with the U.S. Fish and Wildlife Service in Georgia, told us, "We're in the emergency room now."

This is the story of the ranks of biologists, ranchers, ecologists, birders, hunters, wildlife officers, and philanthropists trying to protect the continent's birds from a growing list of lethal threats and pressures. They've created transmitters the size of hearing aids that ride along with migrating birds and relay data back to earth via satellites and cell towers. They're building man-made nests to provide homes for stranded birds and moving entire populations to safer territories. They've borrowed advances from human genome research to study birds at a molecular level. They've discovered how to record bird songs and use artificial intelligence to analyze what's undermining them. Tactics range from community campaigns constructing thousands of tiny wooden houses for bluebirds (the bird at the top of this chapter as well as on the cover) to long-shot ventures like a $10-million-a-year experiment to save Hawai'i's forest birds by neutering deadly mosquitoes.

In addition to scientists, some small private landowners, large corporations, cattle ranchers, and even the U.S. military are embracing new conservation ideas. The lineup includes elements right out of a futuristic fantasy: One nonprofit has landed a multimillion-dollar donation to fund what it calls "de-extinction" that uses cloning to try to bring long-vanquished birds back to life. Whether or not experiments like these succeed, they are already showing specialists how to peer into the genetic interiors of these ancient creatures to try reengineering their genes for the modern world.

Today's birds face a mix of peril and opportunity. The population losses have raised alarms and added urgency—even desperation—to the search for solutions. Scientists hope the stunning loss of billions of breeding birds in North America in the past five decades will be enough to ignite public and political support that's never been easy to build. The nonprofits, research centers, government agencies, and bird groups that have a history of mistrust and competition are recognizing the need to put away their differences and cooperate. "This is a crisis. We're truly running out of time. There can be no tolerance for not working together," says Nadine Lamberski, the chief conservation officer at the San Diego Zoo Wildlife Alliance, a leading research center.

Birds aren't alone in facing threats, of course. Deteriorating ecosystems are affecting all manner of wildlife, fish, insects, and plants. Almost no bird species has been spared, from the most delicate jeweled hummingbirds to scrappy black crows, from a rainbow of warblers to such common species as blackbirds, owls, and sparrows. The loss of birds goes hand in hand with the disappearance of the monarch butterfly as well as bees, insects, and other animals crucial to the balance of nature. But people have always had a deep emotional connection with birds—and their woes help us see the broader loss of biodiversity. The story of birds may be the best way to witness close up—even in our neighborhoods and backyards—the results of a natural system badly out of whack.

Birds provide a list of services that benefit people. They are among the environment's workhorses, pollinating all manner of plant life and acting as nature's farmers in spreading seeds and fertilizing the land. Birds consume an estimated 400 to 500 million tons of bugs a year—a mind-boggling sum when you consider that typical insects weigh just a milligram or two. They help keep water clean, refresh coral reefs, and maintain the population balances of rodents, worms, and snails. Researchers have discovered that watching birds even relieves stress and improves our mental health.

Now they play a new role: As the temperatures and sea levels are rising, birds are real-time barometers of environmental stability—the modern canaries in the coal mine. While the phenomenon of climate change may be far from home and out of view for parts of the country, the plight of birds is a visceral and concrete reminder of the crisis unfolding right in front of us. In that sense, the birds are talking to us—sending out a plea for help. "Birds are an early warning system that nature has provided us," said John Fitzpatrick, director emeritus of the Cornell Lab of Ornithology and the hemisphere's most influential voice of behalf of birds. "They're telling us that we need to look carefully at what's going on."

Birds have long been the most durable form of life on earth. They are among its original inhabitants, descendants of dinosaurs, dating back 100 million years. With their ability to fly and to adapt over time, birds managed to live through waves of cataclysmic disruptions on the planet that wiped out countless other wildlife. Some species have always come and gone with the ebbs and flows of nature. But for most of recorded history, the world's overall bird population remained massive, thought to total hundreds of billions. In North America, as recently as 150 years ago, billions of now extinct Passenger Pigeons—the chicken of that era—would block out the sun when flocks took off at the same time. Clouds of ducks, wading birds, and seabirds swarmed marshlands, oceans, and islands. Bluebirds, robins, cardinals, and jays provided a reassuring choral soundtrack of nature.

Then, over the past century—the equivalent of yesterday against those millennia—large numbers of bird species began to falter as the environments they rely on changed faster than they could adapt. Some species—notably ducks, geese, vireos, and raptors—did hold their own or even grow, and although scientists have long compiled statistics, they didn't yet have the means to track overall bird populations with any precision. That changed in 2019 when a group of researchers announced they'd found a way to calculate the total bird population by blending newly released weather radar archives with a half century of field counts that specialists conduct every spring. Their discovery documented the losses that surprised even veterans who've spent their careers studying birds. "To see it in a single number was an epiphany," says Ken Rosenberg, the lead author of the study that came to be called the Three Billion Bird report published in the journal *Science*. That put the plight of birds at the top of the news with an intensity that had never happened before. Finally, people sat up and took notice.

We Love Birds—Until They Get in Our Way

Our own fascination with birds started a little more than a decade ago while living in Washington, D.C. We were in phase one of retirement, what a friend calls the "Go-Go Years," preceding the "Slow-Go Years" and the "No-Go Years." We'd chosen a lifestyle geared to three Bs: birds, books, and banjos, which Anders has played since high school. One afternoon at a campsite in the mountains of Virginia, Anders was tuning his banjo, plucking four notes, pausing, then four notes more. Suddenly a bird started calling back from the woods in the exact same key and rhythm, almost like a pitch pipe. This duet continued long enough to capture the video enshrined on our website.

We'd both had long careers in journalism. Beverly started as a feature writer and editor, then syndicated columnist, restaurant critic, and cookbook author. Anders was a local news and investigative reporter for twenty years before switching to editing and running newsrooms in Raleigh, Minneapolis, Miami, and Washington. We turned to birding as an antidote to city life and soon found ourselves captivated by the simple wonder of birds: the way a male's feathers transform into vibrant mating plumage in the spring; how birds communicate with songs, chirps, and flips of their wings; the way so many species migrate halfway up and down the hemisphere twice a year guided by the earth's magnetic pull and maps encoded in their DNA.

We learned to distinguish the subtle differences between an Eastern Phoebe and the Eastern Wood-Pewee. (The phoebe has a dark brownish head; the pewee's is gray. If you listen carefully, both birds sing their own names.) We slowly made progress by taking photos each day, and at night we flipped through field guides page by page, comparing our birds to the illustrations. That's how we learned the family groups and similar birds within them—which bird's bill is orange on the bottom and brown on top, which ones have pink legs. In those days you couldn't simply upload the photo to a convenient smartphone app that spits out the ID on demand. But looking back, the time we spent poring over the photos gave us an appreciation for these details. It also morphed into a challenging game. We started chasing birds up and down the Atlantic coast on weekends, our mini travel trailer following behind like a leashed puppy.

As we spent more time in the woods, our fascination veered toward obsession, and we became the butt of family jokes. We also started reading about advances in technology, the evolution of modern research, and how climate change is upending the annual life cycles birds depend on. The more we learned, the more we realized the deep dimensions of the trouble. We started to worry about how this had happened and why these stories weren't being more widely told. When we retired from full-time work a few years ago, we started writing about birds for newspapers and magazines and eventually launched our website, *Flying Lessons. US: What we're learning from the birds.*

We have plenty of company sharing a love of birds. Of all wildlife in the United States, birds attract the largest following. An estimated 50 million people consider themselves birdwatchers, according to a survey by the U.S. Fish and Wildlife Service. Another two million are hunters of waterfowl and other birds, and by paying license fees and taxes on guns and ammunition, they have together built the country's most successful bird conservation empire. When the first phase of the coronavirus pandemic forced people to stay close to home for months at a time, our ranks swelled as millions more people took up birding. Friends started sending us their bird photos to identify right about the time bird feeder sales skyrocketed and birdseed supplies ran short. Just about every news outlet around the country wrote about the newfound appeal of birding. Our favorite headline came from the online magazine *Slate*, which declared: "You Have No Choice but to Become a Backyard Birder."

But that growing appreciation of birds only goes so far. We also got emails and calls from friends asking how to deal with annoying birds at home—House Sparrows belting out a monotonous, irritating, one-note song outside their window or the woodpecker boring a hole in their siding. We reached this unhappy conclusion: Americans love birds—until they get in our way.

That's partly the inevitable result of a human population that's more than 130 times larger than when the United States was founded. As the nation boomed, we harvested the bulk of our old-growth forests, plowed up 60 percent of the continent's grasslands, and drained more than half of our wetlands for homes, farms, and businesses. Along the way, we've so altered the landscape for birds and other wildlife that many cannot find enough food, build nests, or raise their young. What would it take to stop these losses—and how much of that are we willing to do? Is it possible for birds and people to share the same habitats?

We realized the best way to understand what's happening was to go to where the most severe problems are and witness the rescues, research, successes, and failures. These dramatic shifts are a global phenomenon, occurring in similar dimensions around the world. We decided to focus on North America because much of the innovation in research and conservation is happening in the Northern Hemisphere. The plan was to spend a year on the road, mostly living in a twenty-three-foot Airstream. Our model is appropriately called a Flying Cloud, and the trailer makes it possible for us to get up and go wherever the birds are on whatever day that happens to be.

First, we turned the trailer into a mobile office. The dinette got refashioned into a workspace complete with bookshelves and a nook for the printer. We cleared extra clothes out of overhead bins to make room for stacks of manila file folders and office supplies. By the time we pulled out in January 2021, every square foot was filled, and a cooler in the truck contained enough frozen spaghetti sauce to last for a month.

Some of our friends were dubious about how we'd get along in these cramped living arrangements, but it just suits the two of us. There's a queen-size bed, walk-in shower, smart TV with surround sound, two heating systems, recessed lighting, a gas stove and oven, floor-to-ceiling pantry, and two bins just for shoes. The best feature is the nearly 360-degree scenic views. "Oh, now I get it," one friend conceded while taking a tour.

We decided not to mention the inevitable muddy boots, weeks between laundromats, rising before dawn followed by yogurt in the truck, followed by peanut butter sandwiches in the truck, blisters, mosquitoes, ticks, and finally, a dictionary's worth of bird-related acronyms to decipher. "I feel like we're learning a foreign language," Beverly said at one point. We headed first south, stopping in South Carolina, Georgia, and Florida, then west along the lower half of the country to the Southwest and California and back across the northern half. We also traveled to Hawai'i and South America.

In all, our journey covered 25,000 miles, most of it by road, some by air, and more than we'd anticipated on foot through some of the most beautiful and rugged corners of the hemisphere. Best of all, we came face-to-face with birds of nearly every variety, from the most common to some of the rarest on the planet. Here's the story they have to tell.

The Florida Grasshopper Sparrow, the most endangered mainland bird in the United States, is a little-known, nearly invisible species gradually pulling back from near extinction in the prairielands near Disney World.

1

ON THE EDGE OF EXTINCTION

Two secret locations: one in Florida, the other in Louisiana

At first we think we've taken a wrong turn when the street trails off onto a gravel road lined with shacks sagging with age and neglect. But a few minutes later we find a modest sign for the turnoff to White Oak Conservation Center. We pull the Airstream through miles of pines, oaks, marshes, and creeks along the border of Georgia and Florida before coming to a scene at complete odds with these backwoods: A who's who of endangered species is scattered around the clearings. Southern black rhinos huddle in one pen, next to herds of giraffes, gazelles, and zebras, not far from four tigers, two panthers, and dozens of cheetahs. Across the road, cordoned behind steel-and-cable fences, are the last of the Asian elephants retired from the Ringling Bros. and Barnum & Bailey Circus. Off in a far corner of the complex, we find the newly minted generation of Florida Grasshopper Sparrows we've come to see. A wisp of a bird, they seem like an afterthought in the shadow of these marquee species whose origins span the globe. But the sparrows' plight has put them on the center stage for troubled species. They were bred here in captivity and raised in aviaries, each a scaled-down replica of its natural habitat of inland grasslands called the Florida prairie. In a few days, the birds will head off into the real prairie. Their assignment? Find mates, breed, and rebuild the most critically endangered bird species in North America.

This unique subspecies of sparrows is found only in Central Florida in three small clusters not far from Disney World. Few people outside the conservation community even know they exist. Left to their own devices in the wild, Grasshopper Sparrows spend their days hidden away on the prairie's sprawling fields. They're named for their buzz of a song, which sounds more like a grasshopper than a bird. Exceedingly wary, sticking close to the ground, these small, brown birds whose plumage matches their surroundings are all but invisible except on spring mornings just after dawn when the males perch atop reeds and sing their gentle, raspy appeal for mates.

Then in 2003, a routine survey turned up something odd: The group of Florida Grasshopper Sparrows on one of those chunks of prairie went from hundreds to just thirteen birds practically overnight. Then their numbers dropped in the other two core groups as well. Suddenly, the Florida Grasshopper Sparrow was in line to become the next bird on the continent to go extinct. "The population just completely crashed," says Reed Bowman, head of the bird program at the Archbold Biological Station research center that studies Florida's wildlife and plants. "We had no idea why."

Over the next decade, scientists, state and federal wildlife managers, university researchers, foundations, and nonprofits formed a coalition to try to solve the mystery. They checked to see if a disease had swept the prairie, but that wasn't it. Growing ranks of predators, including waves of invasive fire ants, weren't helping matters but didn't explain such a drastic plunge. Finally, so few sparrows remained—just twenty-two breeding pairs altogether—that a single calamity like a hurricane or wildfire could wipe out the entire subspecies. The team decided it had no choice but to take a gamble: Gather up a collection of adult sparrows and try to breed enough chicks in captivity to replenish the wild population. "We had zero experience in little birds," says Steve Shurter, White Oak Conservation Center's director, who worked with endangered species around the world before coming to Florida. "But this sparrow was in real trouble."

For the next five years, the Grasshopper Sparrow's future would hang in the balance at the edge of extinction. Unfortunately, this is where much of the action to save birds is taking place. The United States has built the world's model for protecting depleted species since the passage of the Endangered Species Act exactly fifty years ago. Since then, an assortment of high-profile birds, from Whooping Cranes to Bald Eagles, Ospreys to California Condors, have been rescued. But the pressure on birds is mounting, and the system is showing signs of strain. Scientists are seeing a dramatic increase in not just birds, but fish, mammals, plants, and reptiles that warrant federal protection. Meanwhile, candidates are stacking up, awaiting decisions for years and sometimes decades on whether they'll join the Grasshopper Sparrow on the Endangered Species List. Almost all of the focus on rescuing birds is playing out on the far edge of existence where the costs are highest, the odds of success are longest, and there's little room for error.

Two birds in particular illustrate these dynamics. The Florida Grasshopper Sparrow is the most recent, and the other is one of the longest-running, the Ivory-billed Woodpecker. The sparrow earned its place as the mainland's most endangered bird when it suddenly collapsed. The woodpecker, on the other hand, is a near-mythical bird whose last confirmed sighting came in Louisiana in 1944. While the sparrow has brushed up against extinction, the Ivory-billed Woodpecker has ping-ponged between "gone" and "rediscovered" so many times that its existence is one of the perennial debates in bird circles. The sparrow and the woodpecker sit at opposite ends of the spectrum in everything from their size to their star appeal, but between them they illustrate the complexities, drama, and costs of trying to pull birds back from the edge.

"Everybody Was Jumping for Joy"

The Florida Grasshopper Sparrow's luck may have run low out on the prairie, but the bird hit the conservation jackpot when it landed in the midst of the menagerie at White Oak. As soon as we pull to a stop near the rhinos, a golf cart rolls up and the driver motions us to follow. We'll spend the next three days camping on the edge of the meandering St. Marys River by an Olympic-size pool, just down the road from a dance studio custom-built for Mikhail Baryshnikov. One of the first lessons we learned about how bird species get saved is this: They need champions and millions of dollars in funding. White Oak contributed portions of both, thanks in part to the two billionaires who owned the complex in succession.

The first was the late Howard Gilman. His grandfather founded the Gilman Paper Company in the 1880s, and for a time it was the country's largest privately owned firm of its kind. Shortly after taking control of the family business in 1982, Gilman spent $154 million turning White Oak into a winter retreat with a golf course, lakes, racehorse stables, and homes where he hosted such luminaries as Madonna, Julia Roberts, Al Gore, and John Travolta. Isabella Rossellini, the Italian actress and once the highest-paid model in the world, became a frequent guest after divorcing director Martin Scorsese. Bill Clinton escaped to White Oak to play golf during the height of the Monica Lewinsky scandal. All the while Gilman was amassing a collection of endangered animals. His vision of breeding them for reintroduction to the wild made White Oak a pioneer in the field.

Gilman was a close friend and patron of Baryshnikov, and in 1989 the Russian dancer asked Gilman for help establishing a modern dance company. Soon the world's greatest choreographers were flocking to White Oak. After rehearsals in the studio Gilman built, evening entertainment featured gourmet dinners, a bowling alley, and a speakeasy-style bar. Baryshnikov named his new company White Oak Dance Project, and while in residence a favorite pastime was feeding Gilman's baby giraffes. When Gilman died of a heart attack in 1998, his achievements as a conservationist and his shortcomings as a businessman both became apparent. He'd accumulated $550 million in debt, leaving his entire holdings in disarray. Eventually White Oak went up for sale for $30 million, animals included.

In 2013 White Oak got its second billionaire, Mark Walter, chief executive of the investment firm Guggenheim Partners, owner of the Los Angeles Dodgers, and founder with his wife, Kimbra, of a global nonprofit called Walter Conservation. The two have expanded the White Oak complex to 17,000 acres and built up its research prowess while preserving the many remnants of Gilman's eclectic interests. A ten-foot polar bear that a taxidermist caught in mid-growl guards the lobby of the main lodge. Down a hall displaying imposing photos of Baryshnikov lies a ballroom where the new owners installed museum-worthy dinosaur fossils of a Tyrannosaurus rex and a Triceratops.

The Grasshopper Sparrow wasn't a natural fit for this exotic setting but it didn't take folks at White Oak long to decide to join the project anyway. What the sparrow lacked in glamour it made up for in complexity. Nobody knew why the tiny birds were dying, much less whether captive breeding, a strategy instrumental in saving several other species, would even work.

In a state famous for birds, the case raised questions about wildlife oversight because the last species to go extinct in the United States was also unique to Florida. The Dusky Seaside Sparrow disappeared three decades earlier from the grasslands surrounding Cape Canaveral. State and federal wildlife managers didn't know enough about the Dusky Seaside to create a workable rescue plan and put off captive breeding until it was too late to even try. By 1981 only five of the sparrows remained, all males. "We just couldn't find a female," says Michael Delaney, the state biologist who spent years studying these birds. The last Dusky died in captivity at Disney World in 1987, and the Fish and Wildlife Service declared the species extinct in December 1990.

The Grasshopper Sparrow consortium decided to work with two breeding centers to get started in 2014, White Oak on the northeasternmost edge of the state and Rare Species Conservatory Foundation in South Florida.

Rare Species is a very different place than White Oak. Paul Reillo, a professor at Florida International University, carved his thirty-acre research center from the horse farms and housing developments in western Palm Beach County. Lacking a billionaire's bounty, he built many of the animal pens and aviaries himself. Among Reillo's specialties is breeding parrots from the Caribbean and Latin America, which along with a generous supply of rare monkeys, creates a noisy soundscape. Howls and screeches rise in greeting whenever the tall, rangy professor passes by on the mud-caked ATV he uses to zip around his grounds.

The two centers had to start from scratch trying to breed these birds in what amounted to a laboratory setting. "I reached out to pretty much every program that was out there," says Andrew Schumann, who runs White Oak's bird program. The researchers tried removing eggs from nests on the prairie, but that didn't prove to be effective. In the end they settled on breeding chicks from adults brought in from the wild. Both centers built large, screened cages and covered the floors with clumps of reeds, wiregrass, and sand. White Oak added feeding stations that look like tiny picnic tables complete with sloping roofs to keep out the rain. "This was as close as we could get to creating the prairies," Schumann says.

All the work paid off on May 9, 2016, when the first batch of chicks hatched out of their creamy white and lightly speckled eggs at Rare Species. The tiny bundles of fluff with bulbous eyes were hardly recognizable as birds, but they looked beautiful to Reillo. "They were perfect little sparrows," he says. "Everybody was jumping for joy." Then, about a month later, the chicks that had seemed utterly healthy died suddenly. "They were perfect little birds, and they just fell over dead," says Reillo. The culprit turned out to be a parasite that quickly spread to all the birds at Rare Species. Weeks of debate followed over whether to halt the entire project for fear the parasite could wipe out the last of the wild sparrows if new recruits carried it to the prairie. When the decision was made to move forward anyway, Reillo bowed out.

White Oak faced its own setbacks once the sparrow chicks started to hatch there. The fledglings grew sickly, showing signs of an infection. Eventually the crew figured out the chicks were suffering from a vitamin deficiency. When they added more live bugs caught from nearby fields, the problems vanished. And just like that, White Oak's nursery staff got a new chore for their tiny charges. Every day for two hours they'd walk the fields in the hot sun and sweep up bugs.

Soon afterward chick production progressed to a steady flow, and by the end of the following year, they'd hatched hundreds of healthy sparrows. It was time for the real test: transporting birds to the prairie where a new team of experts would be waiting to see if they could find mates, raise their young, and survive in the real world. "We weren't really sure if they knew how to be sparrows," says Juan Oteyza, a scientist with the Florida Fish & Wildlife Research Institute who oversees the project on the prairie.

An Elaborate Surveillance Operation

Ninety percent of the Florida dry prairie ecosystem is gone, but where it does exist, it looks like an African savanna, a sea of brownish green grasses ringed by distant borders of oaks and palms. The night before the first sparrow releases, Oteyza drops a half dozen of the birds in a mobile aviary filled with grasses, soil, and weeds to mimic the fields where the birds would soon be living. Just after dawn the following day, he unlatches the front side and lowers the metal gate so there's nothing between the sparrows and the prairie. This is the first test of how these pampered, captive-bred creatures will react to the wild. For the longest time, nothing happens. Then one by one, the sparrows hop out into the open and stare at the prairie. Their first reaction is to retreat and each of them hops back into the interior of the aviary. But their glimpse of the wild seems to lure them. First one, then in rapid succession each of the birds hops to the ledge for a few minutes' hesitation, and then dives into the grass. Oteyza is standing thirty feet away, beaming like an expectant father as the last young male bird disappears into the grass. "He's on his own now," he whispers.

In fact, the birds are not leaving humans too far behind on this secret stretch of prairie cordoned off behind locked gates. The round-the-clock care at White Oak will be replaced with an elaborate surveillance operation. Every bird wears an ID number and color-coded tags so the field team, armed with binoculars and spotting scopes, can tell them apart. Some of the birds wear electronic tracking devices to monitor their travels during their first few weeks of freedom. A windowless shed on the edge of the prairie serves as mission control, where Oteyza and his staff prepare for the weeks leading up to the springtime breeding season. The walls are covered in maps marking each male's territory, and pushpins represent every bird the team locates in the field.

In the coming months, the roughly 150 sparrows released in the first experimental season must learn to survive. One wrong move and they'll end up in the claws of a Red-shouldered Hawk or the jaws of a rat snake, a fate so common biologists call the birds "prairie potato chips." This is where the long odds and high costs of endangered-species work mount. More than half of these young birds will indeed succumb to predators. Without those picnic-table feeding stations, others will starve trying to find enough bugs and seeds.

By the time the first wave of sparrows lands on the prairie, the project cost has reached $1.2 million annually, with each bird representing an investment of approximately $10,000. By the time forty sparrows survive to the nesting phase, the investment has climbed to $30,000 per bird. But for those who've worked to put these replacement sparrows in the field, money isn't a consideration. It's impossible to put a value on saving an irreplaceable part of the prairie, says Paul Gray, who's worked on the Florida Grasshopper Sparrow for decades as science coordinator for Audubon Florida, which helps fund the project. "This is our bird," he says. "When it's gone, it's gone forever."

"It's Not Only Working, It's Working Well"

The following spring, we meet up with Sarah Biesemier, the field supervisor for the Grasshopper Sparrow, for her rounds on the prairie that start at dawn every day. She's anxious to scout for clues on how the rescue is working as the grasslands come to life with the first light. A humming stew of biodiversity, among the richest in the United States, sits at our feet in the interwoven fabric of plants, insects, and wildlife that supports hawks, owls, skunks, snakes, foxes, panthers, and a growing number of Grasshopper Sparrows. "There's one there," she says finally. A young male sparrow, gripping a reed thirty yards away, is barely visible through the morning mist. His head is thrown back until he's looking straight up as he makes his plea. This is the sparrow at its most vibrant; his whole tiny body shakes, his beak wide open as he sends forth a high-pitched call for companionship.

The signs Biesemier is looking for appear a little at a time over the spring and early summer. First come the mating songs and antics as the birds get acquainted, and the sparrows raised in captivity prove right from the start that they know the songs as well as their wild counterparts. Then a few birds begin to collect twigs and grasses for nests. When Biesemier hears a change in the morning songs, she knows a bird has gone from looking for a mate to announcing he's found one. The big news comes weeks later when she begins tracking two newly paired birds and discovers from their bands that they're a blended couple: The male is a product of White Oak and the female grew up on the prairie.

This is the busiest and most critical stretch for Biesemier and her team of a half dozen field-workers. They keep careful track of the backgrounds of every bird and spend their days scouting nest locations. Those aren't easy to find. The sparrows choose spots hidden at the base of clumps of grass, covered over to elude predators. As soon as her team finds a nest with eggs in it, they wait until the parents leave to find food. Then comes the race to raise the nest high enough to avoid flooding. Next, they'll surround it with a wire mesh fence to keep out skunks, racoons, and most snakes. They have twenty minutes to complete the transformation, so they've developed the precision of a racing pit crew. Then the wait begins for eggs to hatch to determine if two decades of work will start to rebuild the population.

That progress, too, comes a little at a time. Even though Grasshopper Sparrows are among the most productive of birds, able to hatch two dozen chicks per season, most of their offspring will die. The flock on the prairie produced more than a hundred fledglings that first year, and about half came from the White Oak breeding project. That brought the number of breeding pairs up to thirty-four, and Oteyza, the lead biologist, was still holding his breath. But by the spring of 2022, the project had released five hundred captive-bred birds and the sparrow's prairie population had more than doubled. White Oak remains the chief breeding center, but the project has added three more.

By the end of 2022, the population had grown steadily, and sparrows in the primary research area reached about 120. Most important, the number of breeding pairs more than tripled to fifty. Oteyza can finally see the sparrows veering away from extinction's grasp. "The outlook is so much better than it was just three years ago. It's a changed trajectory that makes all the difference," he says. "It's not only working. It's working very well."

Rediscovering the Nation's Mythical Woodpecker

Around the same time the Florida Grasshopper Sparrow population ran into trouble, another bird was about to grab the attention of the entire country as no species had in generations. On February 11, 2004, in northeast Arkansas, an outdoorsman named Gene Sparling paddled his kayak along a remote bayou in the Cache River National Wildlife Refuge eighty miles east of Little Rock. He happened to stop for a moment to take in the scenery. Sparling was following Bayou DeView, a waterway that passes giant cypress trees, pines, and three-hundred-year-old hardwoods dripping with Spanish moss on its way to the Mississippi River. "I had just set my paddle down and leaned back in my kayak and had the sensation of being the luckiest man on the planet just to be there," he said. That was the moment a large black-and-white bird flew by and landed in a tree sixty feet away. "My God," he said. "That's the biggest Pileated Woodpecker I've ever seen." Then he got a better look at the perched bird.

"I knew immediately this was not a Pileated Woodpecker, and if it wasn't a Pileated Woodpecker there was only one other thing it could be—an Ivory-bill."

The Ivory-bill, as locals call it, is the largest woodpecker in North America. Though it's often mistaken for the slightly smaller Pileated Woodpecker, this is a far different bird. About the size of a crow, it stands twenty inches tall with a nearly three-foot wingspan. Its namesake bill is the color of the keys on an antique Steinway; its eyes are yellow, and the pointed crest of the male blazes fire engine red. An Ivory-bill's unusual call, referred to as a "kent," is often compared to the blast of a toy trumpet. It also drums a tree in a telltale pattern of two powerful strikes, called a double knock, that sounds like a gun going off. It earned the nickname "Lord God Bird" because, according to lore, that's what people cried out the first time they saw one: *Lord God, what a bird.*

21

The Ivory-bill was never a common bird even in healthy populations because a mating pair needs almost ten square miles to find enough food to feed their young and themselves. Before Europeans arrived, its range covered the South, but between the 1800s and World War II, nearly all of these hardwood bottomland swamps and pine forest habitats were lost to logging.

When Sparling got home that day, he posted a note about his sighting on a canoe club's website. A reader familiar with Ivory-billed Woodpeckers noticed it and contacted the Cornell Lab of Ornithology, the country's leading research institution for birds. Even though the woodpeckers hadn't been captured in a clear photo since March 6, 1938, and the last undisputed sighting occurred in April 1944, Sparling's story had a ring of truth to it. So, two Cornell staffers rushed to Arkansas to check for themselves. That began a year-long probe, conducted under strict secrecy to keep the refuge from being overrun with excited birders. The searchers weren't even allowed to tell their families about the mission and got a three-page memo on how to deflect interest without actually lying. In public everyone used the bird's code name, "Elvis." The evidence they came up with wasn't unequivocal, but several sightings, audio recordings, and one blurry 4.5-second video of a bird in flight were enough to convince the team the Ivory-bill did indeed still exist.

On April 28, 2005, Interior Secretary Gail Norton, Agricultural Secretary Mike Johanns, and the Cornell Lab's John Fitzpatrick held a press conference in Washington, D.C., to deliver the news. They announced an unprecedented, multimillion-dollar partnership led by the U.S. Fish and Wildlife Service and the Cornell Lab to search for the woodpecker and to purchase some of the best remaining habitat where it might be.

"I can't begin to tell you how thrilling it is—it's thrilling beyond words," said Fitzpatrick. Later that day he added: "Amazingly, America may have another chance to protect the future of this spectacular bird and the awesome forests in which it lives." In a country still reeling from the 9/11 attacks, a story with a surprising and happy ending was just what people needed.

The media coverage was exhaustive, prompting a bonanza for the small town closest to the wildlife refuge where the woodpecker was thought to be. Brinkley, Arkansas, proudly posted a billboard at the edge of town to greet dozens of scientists and hundreds of search volunteers, along with the reporters, photographers, and birdwatchers from all over the country. Brinkley embraced its status as home of the Ivory-bill and welcomed the instant economic impact. Penny Child's hair salon added a $25 "Ivory-bill Haircut" with close-cropped sides and a strip of hair on top dyed bright red to resemble the woodpecker's crest. Gene De-Priest's restaurant delighted swarms of visitors with an Ivory-bill burger, salad, and dessert. "I've had one group from the Arkansas Audubon Society that stayed in town three days," DePriest told a National Public Radio reporter. "They spent over $3,000 at my place."

Over the next four years the search for the bird would cover a half million acres in ten southern states, from Texas to Florida. Tips of sightings and suggestions poured in, and when they slowed down, The Nature Conservancy and the Arkansas Game and Fish Commission offered a $10,000 reward. When that didn't work, an anonymous donor offered $50,000 to anyone who could provide a video or photo and lead a project scientist to a living Ivory-bill. In the end, after some 20,000 collective hours of searching at a cost of more than $20 million, the best evidence of Ivory-billed Woodpeckers were Cornell's initial recordings and some oval-shaped cavities that indicate a possible nest tree. But no clear photos or crisp videos emerged, and none of the evidence, however hard won, could prove the case without doubt.

The joint push to find the woodpecker gradually wound down with a series of reports, including a 156-page recovery plan the U.S. Fish and Wildlife Service released in 2010. The plan, required by the Endangered Species Act, called for a series of steps to find and protect the bird. But the most significant impact turned out to be the resulting preservation of hardwood bottomlands benefiting all kinds of birds and wildlife that otherwise might have been developed.

When we visit Fitzpatrick in Florida in the winter of 2022, he says he's continued to hear follow-ups on the Ivory-bill, some of which included new evidence. "I'm going to tell you something that has not been published," he volunteers. "We choose not to publish it because, well, it's still sitting out there as a mystery." Just as the clamor was slowing in 2009, a colleague brought Fitzpatrick the longest and clearest recording he'd ever heard from a man chasing after the woodpecker in his boat. "He was racing to catch up with the bird right in the middle of what we called 'the hot zone' out there. It was up in the canopy, so they couldn't get the visual image on it, but he kept the thing going. And you can hear the sound, and the sound that you're hearing is just like ten or fifteen seconds, and those notes are absolutely spot on."

Even so, Fitzpatrick decided the opportunity to prove the bird's existence to a skeptical world had passed.

Cornell and the Fish and Wildlife Service were ready to move on, but a core of true believers was not. Websites and search parties for Ivory-bills persist throughout the South, some of which are little more than fan clubs. But an elite segment of the woodpecker's crusaders are lifelong researchers and scientists, including several of the country's top bird specialists. The Ivory-bill's appeal seems only to grow as the odds of its survival fade.

When we learned that the research wing of the Pittsburgh-based National Aviary and a technology lab at the University of Pittsburgh had teamed up to join the hunt anew, we thought this might be our chance to understand the allure of the Ivory-bill. It took a couple of months of exchanges, but we were invited to join the crew in a secret location in Central Louisiana where they believe the bird is most likely to be found. Every single person on the team had seen the bird at one point or another, they told us. We couldn't help wondering what it would be like to be among them.

On the Trail of the Ivory-Bill

If we have any doubts about how serious this quest is, they evaporate when we pull our trailer into the compound for Project Principalis—a play off the Ivory-bill's Latin name, *Campephilius principalis*. A newly renovated home serves as headquarters, complete with the latest shipment of cameras and audio recorders, a well-stocked kitchen, and maps that stretch in every direction. A revolving cast of a dozen or so volunteers comes and goes depending on the season of the year. During our visit in April 2021, they include Steve Latta, the conservation director at the National Aviary, along with a retired PhD biologist whose career spanned several federal agencies, and a woodpecker enthusiast from South Carolina who spends so much time in the compound she keeps her travel trailer here.

Each person has a story to tell about the time they saw an Ivory-bill. They swap details of the bird's long history like sports fans trading baseball stats. One minute, they laugh about the "Bigfoot and Sasquatch" comparisons their search inevitably evokes. The next minute, they turn deadly serious. Latta, a busy scientist with a wife and kids at home, talks about why his sighting led him to join the hunt and align the National Aviary with the cause. "It's real. It's out there," he says. "And I just feel this huge responsibility. This is what motivates us. The weight of this thing. That's what keeps us going."

Early the next morning the crew loads gear, and we drive out to one of the target tracts. The location is confidential for several reasons. When word gets out that one of the country's most iconic endangered species might be nearby, not everybody is pleased. A few years ago, Mark Michaels, a lawyer from New York and a Principalis founder, was looking for the woodpecker near a large section of privately owned forest. "They got wind of it," he says, which apparently provoked fears the owners would lose control of their land if the bird was found there. "Oh my God, they just went in and mowed everything down," Michaels says.

Another reason for secrecy is to avoid tipping anyone off about their progress. If Project Principalis seems to be searching for buried treasure, they want to be the ones to find it when and if the time comes.

We drive for miles before the turnoff that takes us deep into the woods and the target area. Finally, we park, grab cameras, backpacks of provisions, and start along a mushy bottomland in the middle of nowhere. This is where the Ivory-bills of long ago liked to nest, according to the late James Tanner, whose firsthand accounts of the Ivory-bill in 1937 and 1938 became his doctoral thesis and remain the definitive authority on the woodpecker today. It's easy to imagine how even the largest woodpecker could disappear in a forest so thick with mature hardwoods, cypress, downed trees, swamps, and waterways. It's nearly impassable in places, and Latta warns us to keep an eye out for anything that moves, especially the venomous water moccasins. He's wearing thick hip waders as protection. "It's usually the third person in the line that's in danger," he says. (We're wearing standard hiking boots, and so we quickly trade places because one of us is terrified of snakes.)

The hike goes on for hours, and there's plenty of time to hear why Project Principalis is betting it can find a bird that half of the nation's experts, along with a whole generation of amateurs, have been looking for since 1944. The technology has finally reached the point where motion-detection cameras, audio recorders, and drones can help zero in on the woodpeckers' range, and then locate a cavity where the bird might be nesting. That's the goal—since the woodpecker must be reproducing for it to have a future. The Principalis team has posted recorders throughout the most promising tracts, and one member uses drones to scan the woods. When they find any evidence an Ivory-bill might have nested or foraged nearby, they set up motion detection cameras to record any activity. They also collect DNA from anywhere they think the bird has been, though they've yet to get a match.

As we trudge through the muck, under towering trees, everyone talks about how they became converts. Latta, a PhD ecologist who's traveled the world for the aviary working on conservation projects, was intrigued when he learned that a University of Pittsburgh bioacoustics lab planned to participate. He was also skeptical, but then on his first visit here four years ago, he went walking in the woods with Michaels. The two had stopped for a break when a flash of movement caught Latta's eye.

"I don't know where it came from, but it came off pretty low, and it was flying up, up, up for five, six, seven, or eight seconds," Latta says. "I got this full view of the back of the bird and—this is embarrassing but it's also the most telling part—I'm not the greatest birdwatcher, but I knew enough to look for the white on the trailing edge."

Latta is describing how an Ivory-bill's wings are white on the edges in flight, whereas those of the Pileated Woodpecker are black. This is the main point of controversy over which woodpecker was captured in the 2004 video from Arkansas. "I saw white, and I thought, 'White on the trailing edge,'" Latta says. "I'm not seeing what I wanted to see. I'm seeing what I saw, and it matches exactly." He stood for a time in the woods just trembling. Then he and Michaels, who didn't see the bird this time, spent an hour circling the area in vain.

Later that evening, when Latta comes back to his story, he suddenly tears up as he talks about his realization the bird still existed. "The thing that shocked me was that for several nights after that, I was like: 'It's real.'"

Of all the volunteers, Tommy Michot is the most familiar with Louisiana. He's a scientist who worked for federal agencies and universities here for forty years, including a time with the U.S. Fish and Wildlife Service. He's seen the woodpecker twice, he says, twenty-six years apart. His best look was in 1981, when he and a coworker had such a close view they could see the precise telltale white wing edges. "It flew at an angle so we had a much longer look," Michot says. "We both almost had the exact same words for what we saw. It was like a flying cross and it was long."

South Carolina volunteer Peggy Shrum also got a lengthy look at the bird she's been dreaming about for decades. These days seeing an Ivory-bill opens you to ridicule and outright derision, she says. "I don't typically expect people to believe me, but I don't care if they believe me," she says. "I know what I saw. I have no doubt at all."

A Pot of Gumbo and Scanning for Evidence

At the end of a full day in the woods, the most important part of the work begins. After a break to shower off the mud and sweat, everyone gathers around a computer to page through the photos and sounds downloaded from the recorders and cameras positioned in the woods. Michot heads up the review with a notebook on one side and a glass of red wine on the other, while a pot of homemade gumbo simmers on the stove. He narrates what the motion-triggered cameras have picked up from the past several weeks.

"There's a wild boar. That's a Red-shouldered Hawk. There's a duck." He quickly clicks through the pages. "A deer. That's a rodent of some kind. There's a coyote. Squirrel. Another squirrel." He explains why the camera is aimed at a fallen tree. "Yeah, that's the target tree. We thought maybe it had been scaled [marked] by woodpeckers. You know, the bark feels about the right stage of decay. We thought it might be . . ." He trails off.

Then a photo pops up with the closest likeness we'll get on this day. It's the smaller Pileated Woodpecker, which shares the Ivory-bill's black-and-white plumage. The photo is far sharper and more tightly focused than most of the shots and would have made for convincing evidence. "Imagine if that was it," Michot says. "It's pretty darn good."

While the group pages through the photos, a bioacoustics lab at the University of Pittsburgh is busy analyzing hundreds of thousands of hours of audio recordings Principalis has collected in Louisiana over the past two years. The recorders are programmed to vacuum up sounds from the forest during the birds' most vocal stretches in the morning.

While the concept is simple, turning the recordings into concrete leads is time-consuming. The only good-quality audio of an Ivory-bill was recorded eighty years ago with antiquated equipment when the woodpeckers were agitated. The recording isn't very clear, so it's hard to convert into a digital graphic that can be used to search for sound matches. "That's turned out to be more difficult than I think any of us thought it would be," Latta says. "So we're not there yet."

The following morning we hitch up the Airstream and point the truck west. Although we didn't really expect to see the "Lord God Bird" in person, we did experience its allure secondhand. Just being in the eerie woods and hearing the stories from our generous hosts was enough to offer hope. And they aren't alone in reporting continued sightings from all over the South. Since the last confirmed view in 1944, some two hundred reports have included video and audio recordings, but most reports are firsthand accounts from sportsmen, ornithologists, biologists, and ecologists.

In all that time, not one unequivocal photo or video has turned up, which is a good part of the reason for an announcement by the U.S. Fish and Wildlife Service on September 29, 2021, that set off alarms all across the woodpecker fandom. The agency completed a review and said it was finally time to give up and declare the Ivory-billed Woodpecker officially extinct. Amy Trahan, a biologist with the service who put together the report, said she had no choice but to recommend removing the bird from the Endangered Species List. "That was probably one of the hardest things I've done in my career," she said. "I literally cried."

The Ivory-bill provoked strong reactions when it was considered alive, and that didn't stop with an extinction declaration pending. An outpouring of objections came from scientists, bird organizations, and others who'd searched on their own. Along with the woodpecker, twenty-two other birds, including eight from Hawai'i, were to be stricken from the list. Only the Ivory-bill provoked objections and got a public comment hearing. The hearing didn't come off as a genuine inquiry, however. Fish and Wildlife used an outside organization to manage the video event whose leaders seemed to know nothing about woodpeckers.

Many who signed up to testify weren't aware of the two-minute limit, which meant quite a few got shut down in mid-sentence. That included John Fitzpatrick, one of the first to argue that the decision was premature and served no good purpose. He was just getting wound up when the organizers cut off the very scientist who'd announced the Ivory-bill's return at the national press conference in 2005. "The hearing was a joke. It was a total joke," he says later. "I'm not saying the bird is not extinct. I'm saying it *might* not be extinct. And if it might not be extinct, then don't delist it yet. It's serving a purpose. It's helping to protect this habitat. It's also so iconic in lots of different ways. So what's the rush?"

The most detailed response to the service's proposal came from Project Principalis. Led by Steve Latta and Mark Michaels, the group published a thirty-eight-page academic paper in April of 2022 summarizing their findings under the title, "Multiple lines of evidence indicate survival of the Ivory-billed Woodpecker in Louisiana." The report walks through the details of the many sightings by Principalis volunteers and includes videos and page after page of photos, although none clear enough to settle the question. "As we put this together," says Michaels, "it was the impact of that cumulative body of evidence that was particularly powerful to us." The response to the prospect of declaring the bird extinct was so extensive that Fish and Wildlife announced in July 2022 they planned to delay the final decision on extinction for at least another six months. The statement made it clear that the one thing the service needs to change course is precisely what nobody has produced in a century: a clear and convincing photo of this "Lord God Bird."

Why We Love Long-shot Rescues

Dave Naugle, a wildlife biology professor at the University of Montana, has a theory about why Americans love it when birds are plucked from the jaws of extinction at the last moment. "It tugs at our heartstrings," he says. "It's like, 'Oh my God, there's just one left. You've got my attention.' This is something we can win."

The country's focus on long-shot rescues goes back to the beginning of the Endangered Species Act (ESA) with the recovery of such high-profile birds as the Bald Eagle, Peregrine Falcon, and Osprey. The 1973 legislation is the only guaranteed source for rescue funding and also the only clear mandate for action. Almost every bird featured in these pages is either federally listed or perilously close. At the start of 2023, there were about a hundred birds designated as endangered or threatened under the Endangered Species Act, and this is where much of the action and the innovation in North America's recovery of species takes place.

But the last-minute approach is growing increasingly ill-suited for a time when so many birds in North America are losing ground. The longer it takes to respond to a failing species, the more expensive and complex it becomes. "The way that public money has been spent has been to put out fires, and that's been important," says Scott Sillett, head of the Smithsonian Migratory Bird Center. "But the reality is that putting out a fire is less effective than preventing a fire from starting and figuring out how you proactively manage these populations." The fire at the opposite end of early intervention will mean even more nests holding baby sparrows worth $30,000 apiece.

This was the backdrop when a group of researchers struggling to get the public's attention went to work digging into the complex, confusing, and shifting outlook for North America's birds. What they found shocked even the most experienced scientists among them.

The Eastern Meadowlark, whose melodious song is part of the soundtrack of the nation's grasslands, has lost 75 percent of its population to development and farmland over the past half century.

2

VANISHING BY THE BILLIONS

Ottawa, Canada

For weeks, Adam Smith had been crunching the raw data from more bird statistics than anyone had ever tried before—thirteen different bird counts and millions of radar sweeps. Suddenly he heard the musical chime that tells him his results are ready. He leaned across his desk, surrounded by enough high-powered computers to heat up his entire office, and stared at what could only be an impossible conclusion: Over the past fifty years, his calculations found, a third of North America's birds had vanished. "Well, that can't be right," he thought. "I must have made a mistake somewhere."

Smith, one of the hemisphere's top specialists in bird populations, just sat for a while in his cluttered space at the Canadian Wildlife Service, which was decorated with caribou antlers, a musk-ox skull, and early drawings from his twin boys. Then it dawned on him. "This would be a massive change, an absolutely profound change in the natural system," he said. "And we weren't even aware of it."

Up until that point, counting the abundance of individual birds throughout the entire continent was impossible. At any given time, many species number in the tens of millions in North America—adding up to billions of birds—and they're constantly on the move. But the science of bird study was advancing, and a close-knit group of scientists was experimenting with using radar imagery, satellite photos, and citizen science to add precision to the dozens of conventional bird counts done for groups of species. The computation Smith had just finished that day in May 2019 combined individual population estimates for 529 bird species, from the most common sparrows and robins to rarities hardly ever seen. When Smith pulled these estimates together and adjusted each for its degree of certainty, the findings came down to a single ski slope of a chart. It showed a precipitous drop in nearly all these species in every part of the continent. At the bottom sat four lone digits—2.913. That's the number of breeding birds in billions that had disappeared since the early 1970s. He had documented an accelerating churn of seasonal losses that had slowly taken their toll on the abundance of birds. And it translated to just short of a third of the adult birds that not long ago filled North America but now were gone.

The hardest hit were grassland birds, down by more than 50 percent, mostly due to the expansion of farms that turn a varied landscape into acres of neat, plowed rows. That equates to 750 million birds, from bright yellow Eastern and Western Meadowlarks with their incessant morning songs to stately Horned Larks with black masks across the male's eyes and tiny hornlike feathers that sometimes stick up from their heads. Forest birds lost a third of their numbers, or 500 million, including the compact, colorful warblers and speckle-breasted Wood Thrushes that sing like flutes. Common backyard birds experienced a seismic decline. That's where the bulk of the loss of abundance occurred, among just twelve families of the best-known birds—including sparrows, blackbirds, starlings, and finches. There's been relatively little research on these species, and there's no sense of urgency when resources are already stretched thin for so many other birds.

The possibility of such losses was too startling to share with his colleagues until Smith checked every step of his calculations, particularly since he'd never attempted this analysis before. "It always takes a couple of times to get these numbers right," he says. After a day and a half of painstaking scrutiny, Smith realized there was no mistake. "I was speechless. We'd lost almost 30 percent of an entire class of organisms in less than the span of a human lifetime, and we didn't know it."

Just One Question; Just One Number

Ithaca, New York

Adam Smith's quest for the one big number got its start two years earlier when some of the best minds in bird research gathered at the Cornell Lab of Ornithology in the small upstate New York town of Ithaca. The 90,000-square-foot research center is a global mecca for the study of birds appropriately situated in Sapsucker Woods, an unspoiled sanctuary at the edge of a pond. Nearly everything about the lab is designed around birds, from the abstract birdlike shape of the building to its three-story windows offering the best possible views of the surrounding woods rich with birdlife. It only takes one unusual species to bring even the most serious meeting to a standstill, as everyone lunges for the binoculars waiting along every windowsill.

The lab is also a draw for ordinary birders who flock here from all over the world, reverently lining up for tours or spending time at the many scopes set up throughout the two-story, paned-glass lobby. Whether you're entering the building for a tour or coming to research bird populations, the first thing you see is a floor-to-ceiling mural of all the world's current and extinct bird families. Behind a locked door on the first floor, seven-foot-tall towers of drawers hold more than 60,000 specimens of existing and extinct species collected from around the world dating back to 1831. Down the hall, the K. Lisa Yang Center for Conservation Bioacoustics has gathered so many digital graphs of bird songs—millions of files that capture sounds for two thirds of the world's roughly 11,000 species—that the lab is always looking for the latest cloud solutions to hold them all. The lab's Macaulay Library houses 38 million bird photos, 235,000 videos, and a million bird song recordings. The lab's research, images, and data have helped make birds the most studied slice of wildlife on earth.

But when three dozen scientists from research centers, federal agencies, and nonprofits in the United States and Canada gathered at Cornell in July 2017, the focus was on what they didn't know. The scientists meet each year as part of a cooperative nonprofit called Partners in Flight that pulls together bird researchers to tackle a running list of scientific quandaries. For years they'd been working on how to make the most of rapid advances in a profession that's hardly recognizable from when many of them began their careers. In those early days, research was largely limited to what ornithologists could see, hear, and jot down while observing birds in their natural environments.

Now their ranks include ecologists, biostatisticians, computer scientists, and acoustic engineers using fresh technologies for tracking and studying birds and bringing together evidence of dramatic shifts in the avian landscape. Partners in Flight built a color-coded watch list of species in varying degrees of trouble. During that July conference, the prestigious core group known as the Partners in Flight Science Committee spent days struggling to refine the ways they count individuals in each specific species. It's not enough to know how many sparrows there are, for example, because more than forty varieties live in North America—White-throated, Chipping, Lincoln's, and so on—each with a distinct place in the natural kingdom.

Then toward the end of the third day, the discussion took an abrupt turn with just one question, posed by an outsider—a nonprofit executive without any sort of degree in bird biology. "Exactly how many birds have we lost altogether?" Mike Parr asked. "Can we put a precise number on it?"

Parr had just been promoted to president of the American Bird Conservancy (ABC), a nonprofit based in the suburbs of Washington, D.C., that funds research, promotes on-the-ground conservation of the hemisphere's most threatened birds, and lobbies Congress. Parr was invited to stop in on the gathering by Ken Rosenberg, a conservation scientist and the country's leading expert in bird populations whose position was jointly funded by ABC and the Cornell Lab. "Mike was really pushing us," recalls Rosenberg. "Were there actually fewer birds today? We didn't know the answer to that."

A Bird Universe Sprawling Everywhere

Birds exist in a sprawling parallel universe that's mostly hidden from view. Over millions of years, they've evolved to form elaborate societies spread across every habitat on the planet: forests, grasslands, wetlands, coasts, the open seas. Some birds stay put while others travel along specific routes, airborne highways really, up and down the hemisphere, following shifts in weather and food. These rough corridors that we call flyways stretch north to south from arctic Canada to the bottom of South America. Birds' daily routines take place mostly in seclusion as they follow precise schedules for courting, mating, raising young, and teaching them the ropes before sending their fledglings off to fend for themselves. Every fall and spring, the migrators—some 350 of the roughly 1,100 species in North America (mainly songbirds, ducks, gulls, herons, and other shorebirds)—take flight for the transition from summer breeding grounds in cooler, more northern territories to points south where they wait out the winter.

Several compelling, bestselling books have recently brought to life the discoveries found in the advances in bird study. Contrary to the "bird-brained" label, the latest research has detected uncommon intelligence inside their tiny skulls. Crows can solve puzzles, give gifts, and remember where they've hidden thousands of seeds stored for winter. Birds have highly developed communication skills, fondness for play, and show acts of kindness as well as outright deception. Their flight skills are astounding. Some birds use GPS-like instincts to return to the same destination year after year. Even the love lives of birds have turned out to be much more intriguing than once thought. Elaborate plumage and dance rituals of such birds as peacocks and Birds of Paradise reveal an appreciation for beauty and aesthetics in their romantic relationships.

Along with the study of how birds live, researchers are now increasingly focused on how they die in hopes of helping to protect them. This, too, is a complicated arena, loaded with guesswork and rough estimates. A huge percentage of birds have always succumbed to natural causes in the bird-eat-bird world they're hatched into. Depending on the species, as many as 90 percent of chicks don't survive to adulthood, doomed by everything from hungry predators to starvation to turbulent weather. Even so, by the fall when a new generation takes flight after the spring and summer breeding season, scientists estimate North America's bird population swells to between 30 and 40 billion. By the following spring at the start of the next mating season, the total number of breeding birds reaches an equilibrium, thought for the past fifty years to be roughly 10 billion. This is the number against which the newly documented three billion losses are measured. The new normal for the start of breeding season is down to roughly seven billion birds. No matter how you slice it, that's a lot of lost birds.

Throughout tens of thousands of years, even millions in some cases, evolution has shaped everything from wingspans and beaks to dietary habits and migration. The five-inch-long Golden-cheeked Warbler, for example, can only build its nest with shreds of bark from Ashe juniper trees. These are becoming harder to find in the warbler's Central

Texas breeding grounds near urbanized Austin, where 97 percent of land is privately owned and where most junipers have been cleared for homes, farms, and businesses. Many other birds—such as the Florida Scrub-Jay and the Red-cockaded Woodpecker—evolved to be dependent on seasonal fires set by lightning. Now such fires are rare in much of the continent, and the public opposes controlled burning, especially in drought-stricken areas prone to wildfires.

Finally, a laundry list of man-made hazards compound nature's culling. A major study by the National Audubon Society documented how climate change is shifting birds' territories and threatening their survival. Sea level rise is flooding nests of some ground birds breeding along coastlines and tidal waters. Collisions with high-rise glass-covered buildings, power lines, and vehicles kill more than a billion birds a year. Outdoor cats kill an average of 2.6 billion birds every year in the United States and Canada mostly for sport. In still preliminary estimates, wind turbines are thought to kill as many as a million birds annually, with wind power expected to undergo a dramatic increase in the near future.

Pesticides, and in particular a newer class of insecticides called neonicotinoids that are the most widespread in the United States, are proving more harmful to many birds than was once thought. But the Environmental Protection Agency, which is responsible for assessing the dangers of pesticides, admitted this past year that it has fallen far behind in assessing what these chemicals do to wildlife and cannot say what the true hazards of many common household and agricultural pesticides are. That has meant that no one is sure what role pesticides may be playing in the decline of bird populations, although there's little question they are a contributing factor.

All in all, it's a complicated mosaic that scientists haven't been able to communicate effectively to engage, or perhaps even enrage, a broad audience. These were issues on Mike Parr's mind when he arrived at the Cornell Lab that day in 2017. He decided this gathering of the nation's top scientists would be the right place and the right time to take a different approach to making the case that something utterly perilous was happening to birds.

How Long Can Birds Wait?

Mike Parr has never been shy about speaking his mind. Born in Liverpool about the time the Beatles rose to fame, he was captivated in primary school by the birds in his neighborhood, an instant fascination that has endured to this day. Among his prized possessions as a youngster were collections of old bird eggs and stuffed hummingbirds his parents found at a flea market. Just a few years later, he was hounding his teachers to take him birding at nearby preserves. "It was my sole, complete obsession by the age of fifteen," he says during an afternoon of birding in a park outside Washington. If you're talking with Parr anywhere and he hears a bird, throwing out the ID is his reflex response. "There's a titmouse," he says, then picks up right where he left off as we walk the trails.

Before college he set off alone to travel the world and chase birds. Unlike most of his colleagues today, Parr was never drawn to the advanced degrees common in the field. Instead, he made his way on his persistent curiosity and outgoing nature, qualities that helped him land his first job in the field with the global nonprofit BirdLife International in Cambridge, England, which eventually led him to the United States and the American Bird Conservancy. Parr worked for the organization for twenty years, finally as head of programs and communications. Since 2017 when Parr took the helm as ABC's second president, the annual budget has doubled. "It's great to have a load of ideas but if you can't pay for them, they remain on the drawing board," he says. Parr decided bird nonprofits needed to do more to communicate their mission. Despite all the knowledge collected on birds, Parr believed that he and his peers offered a muddled picture of their status.

"I felt that we as a community hadn't communicated the seriousness of the problem," Parr says. "People would say, 'What's going on with birds?' And the answer would be, well, some are declining, some aren't, some are about the same. By the time you've said all that, you've lost everybody. We did know, we just hadn't figured out how to answer the question properly."

Parr also worries that scientists—particularly the academic segment that controls much of bird research—seem content to study the problems for too long before taking action. "We can't afford to wait for perfect science to do things. If we just study things, we'll study them into extinction. So, I say let's do what we think is best now." Parr hoped coming up with the total change in bird population for the first time would make a difference. "I thought one way to make the case would simply be to figure out the essence of the problem. How many birds do we have versus how many we had before?"

Parr's perspective wasn't all that welcomed that day at Cornell. Tom Will, a longtime Fish and Wildlife biologist in the division of migratory birds, picked up some resentment when Parr started talking. "A number of us in that room were skeptical," Will says. "Here's Mike giving us a challenge, you know, while we've been around for twenty years." But Parr's question had already crossed the minds of others in the room, and a couple of the statisticians among them had done some preliminary work on calculating broad population numbers.

The challenge of calculating an exact number got them all talking. "I remember somebody said, 'I think we've got fewer,'" Parr recalls. "And a couple of other people said they weren't sure. I think somebody else said that it's about the same, but they weren't sure either." Adriaan Dokter, one of the world's experts on the use of radar to study birds, said the challenge was as basic as it was compelling: "It's such a simple question that you think, okay, we must know that. But nobody ever looked at that. Overall, have the birds declined or not?"

When the day ended, the group agreed to see if they could indeed come up with a definitive tally on 529 species that represent the core of the North American breeding bird landscape. The first step was to collect all the existing data, which was spread among a dozen different censuses, some done by regions, and others broken down by species types such as ducks, raptors, songbirds, and shorebirds, much of it collected over decades of counts. The data was so extensive, the project was a perfect fit for the big data computing that's developed in parallel fields of business, engineering, and medicine to mine a wide landscape of data and make calculations.

Then they used a secret sauce to check the findings with a new technology more precise than the annual bird counts: weather radar records that can be filtered to count birds as they fly. The radar, recorded by 143 state-of-the-art stations around the United States, is designed to capture high- and low-pressure zones, rainfall amounts, and wind speed. But images of birds show up, too, even at night when much of the migration flights occur. After the last decade's worth of archival records were released to the public in digital form for the first time in 2015, the data added a new level of confidence to counts for the most recent years. Dokter, the global radar expert now working at the Cornell Lab, spent months layering in the additional data. "It matched perfectly," he says.

In the spring of 2019, Adam Smith, the Canadian biostatistician whose role it was to pull all the data sources together for analysis, finally heard that chime on his computer with the results. He sent out an email with two graphics attached. One showed the historical bird populations for each of the major land types on the continent, with a slight increase for wetland birds and declining numbers for nearly everything else. Then he built a single graph revealing that the collective losses totaled 2.9 billion birds. "It was a eureka moment," he says.

The results blindsided even those who'd tracked populations their entire professional lives. Dozens of back-and-forth emails from that year provide a glimpse of the complexity of this undertaking. The

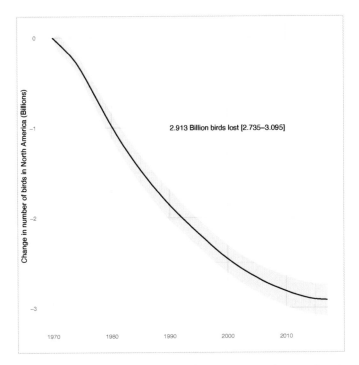

The graphic that Adam Smith sent out to his fellow scientists, showing the seismic drop in bird populations over the past fifty years.

messages run for pages with scientific discussions about how to make the data foolproof. And every so often there are comments showing the best minds of the bird world pausing to contemplate the magnitude of the grim evidence taking shape. "Very shocking," wrote Dokter in an email at one point. "Sobering," Mike Parr messaged the entire group. "Extremely sobering."

"A Defining Environmental Challenge"

As soon as Pete Marra heard about the scale of the losses, he got on the phone to Ken Rosenberg. The two had met in college and became close friends over the years. "We've got to get this into *Science*," Marra said. He was talking about the premier academic journal and by far the hardest from which to win acceptance. "It's going to change the way we look at this," he insisted.

Marra wasn't part of the Cornell gathering, but he was head of the Smithsonian Migratory Bird Center in Washington, D.C., and had worked with birds his entire career. He immediately pitched in with Rosenberg on the next step: turning these numbers into a narrative for the journal in order to present their findings in a venue that would demand public attention. "It's important to get this published where it has credibility," he said.

Marra is a workhorse, which dates back to his upbringing in a broken home. He was raised in part by a harsh grandfather for whom he's thankful for at least one thing: "He did teach me to work hard." Marra escaped from his tumultuous home life to a nearby nature center, where he fell for birds and began taking on research assignments in the summer throughout his school years. By the time he finished his PhD, Marra was in demand as a researcher and he quickly landed a job at the Smithsonian. In 2019, while working with Rosenberg on their paper, he joined Georgetown University as director of the university's new school on the environment.

The day we visit Marra at Georgetown, he's just returned from a work trip to Greece and will soon be headed off for meetings in Western Canada. Deep circles under his eyes betray the pace he keeps—so hectic that it worries his close friends. But Marra says he's happiest when he's busy like this. "I never sleep much," he says. "I just want to be working." The evidence is spread around his office. Photos of his worldwide research trips decorate the walls. Framed covers of stories he's written for academic journals run across one side of the room like trophies.

As a draft for *Science* came together, Marra and Rosenberg added a rich library of background data, graphs, charts, and spreadsheets to anticipate the questions the report was likely to face. The report was a blend of scientific terms and sweeping conclusions. "Slowing the loss of biodiversity is one of the defining environmental challenges of the 21st Century," they wrote.

Once the editors at *Science* agreed to publish the Three Billion Bird paper, a new question arose: What would it take to grab—and keep—public attention? The importance and difficulty of capitalizing on the moment was obvious to everyone. So, in the months before the release of the *Science* report, the expanded team of contributors turned to a PhD biologist who'd switched to a career in science writing at the Cornell Lab. If anyone could figure out how to make the message stick, they decided, it would be Miyoko Chu.

Hashtag #BringBirdsBack

There weren't many birds in the San Francisco neighborhood where Chu grew up. But one day her family traveled to the city's Chinatown, where they came upon poultry trucks selling live pigeons for food. "I started to cry," Chu remembers. "Right on the spot, my dad, who had raised pigeons as a boy in Shanghai, forked out $2 apiece for several of them." He tied up their feet, Chu says, put them in a bag, went home, and built a pigeon coop.

Chu didn't watch her birds from a distance. She climbed right in and sidled up to them. "So I used to sit inside the coop. They were not tame in the beginning, but over time they came to really just trust me, as if I was another pigeon. And so I could see everything close up, all the courting and the mating, the egg laying, raising their young. It was a complete front-row seat, and I could even reach in there and touch the nestlings. That was how my interest in birds started."

Chu set out to be a biologist, working her way through grad school and eventually completing a PhD. But she had always loved writing about birds as much as researching them. When she started a family, the demands of field-work—from rattlesnake and bobcat encounters to long days in the wilderness—pushed her toward writing. She joined

the Cornell Lab and eventually took over its communications department. In a field where scientific jargon is an occupational hazard, Chu
is a thoughtful interpreter for the lab. But the all-out effort the *Science*
paper would require was something else altogether, and the stakes had
never been higher. Here was a chance to finally do something else that
had never been done before: get people to care.

During the year her colleagues were perfecting the *Science* paper,
Chu pulled together a team to figure out the best way to get a broader
audience to understand the plight of birds and explain the impact birds
have on people in their daily lives. For years, the Cornell staff believed
that the birds themselves could make that case. That may be true for
a segment of the public, but most people don't pay much attention to
birds and know little about them.

In the end, Chu's team let the numbers do the talking. They built
a website called 3Billionbirds.org. Click and you see a meadowlark dis-
solving in a cascade of feathers and the headline "3 Billion Birds Gone."
Underneath that: "Together we can bring them back." And then the
kicker: "Birds are telling us we must act now to ensure our planet can
sustain wildlife and people."

When *Science* published the report on September 19, 2019, the
news ran on front pages around the country, at the top of newscasts
on television and cable networks, on radio stations and throughout social media. The headlines were bold and succinct. "Birds Are Vanishing
from North America," said *The New York Times*, with the first sentence,
"The skies are emptying out." NBC News reported: "U.S. and Canada
have lost 3 billion birds since 1970. Scientists say 'Nature is Unraveling.'" Overall, the campaign got more than 25 million search results on
Google.

Gradually over the course of the announcement day, it became clear
just how broad the reach was turning out to be. Mike Parr was in Monterey, California, for an American Bird Conservancy board meeting,
and midway through the afternoon, he stepped out to do an interview

with a foreign publication. He saw Ken Rosenberg pull up, and when he opened the door, he realized that Rosenberg's voice was on the car radio in an interview. "Ken was driving the car, and he was on the radio at the same time," Parr says. "I'm like, 'Oh, this is big.'" Parr lingered outside, intently watching the bay, when a man walked by and paused. In his typical outgoing fashion, Parr started up a conversation and pointed to some birds a few yards away. "I said, 'Those are Brandt's Cormorants.' And the guy said, 'Yeah, the birds. I was just hearing about this today. It was in *The New York Times*. There's a big problem with birds, isn't there?' And then I knew. We'd broken through."

Another *Silent Spring*

This isn't the first time the future of North America's birds have been cast into jeopardy. In 1962, Rachel Carson published *Silent Spring*, her landmark book that documented the devastating impacts of the overuse of pesticides on birds like eagles and pelicans. A shy, gentle introvert with thick black glasses and a background in marine biology, Carson spent five years building her case: that the nation was poisoning its wildlife, plants, soil, and people to rid itself of insects. "How could intelligent beings seek to control a few unwanted species by a method that contaminated the entire environment?" she wrote. "But this is precisely what we have done." Carson used birds as a core messenger in her book, whose title referred to the silencing of bird songs in a portrait of desolation she painted in the opening chapter. Serialized in *The New Yorker*, *Silent Spring* was a bestseller before it even reached bookstores, prompting a flood of outrage from readers—as well as criticism.

The chemical industry in particular vilified and parodied Carson. But she had spent years as a biologist for the U.S. Fish and Wildlife Service and had utter command of the science—plus a level of fortitude unknown publicly at the time. While she was pushing to finish her book and then taking on public demands, including nationwide speeches and testimony before Congress, Carson was dying of breast cancer. She passed away two years after her book came out, and never knew the lasting impact of her work. It led to the banning of dichloro-diphenyl-trichloroethane, or DDT, the insecticide with the most impact on birds. The Environmental Protection Agency and Endangered Species Act were both put into place in the early 1970s in response to her book, which also helped ignite the modern environmental movement.

Now, scientists behind the Three Billion Bird discovery are hoping their study will prompt a similar response. "I do think this could be the next *Silent Spring*," says Marra. Ken Rosenberg goes further: "*Silent Spring* changed the playing field. It changed the way people thought about conservation. That's what Three Billion Birds is actually doing." The jury's still out. The forces working against birds are more varied and complex today, starting with political divisions that make anything like the congressional action that followed *Silent Spring* unlikely. On the other hand, change didn't come quickly back then, either. It took another eleven years for Congress to pass the Endangered Species Act and eight years to launch the Environmental Protection Agency.

The infrastructure devoted to birds is also far more extensive today. The Fish and Wildlife Service, headquartered in a marble, fortress-like building off Constitution Avenue, has a staff of 9,000 around the country overseeing America's wildlife. With a massive load to manage and not nearly enough funding, the Fish and Wildlife Service is a big ship to steer. The service, housed under the Interior Department, is one of nine federal agencies with substantial roles in helping or researching birds, including the Departments of Defense and Agriculture. All fifty states each have their own wildlife agencies as well. There are four major nonprofits concentrating on birds—Audubon, the Cornell Lab, American Bird Conservancy, and Ducks Unlimited. Audubon has the largest membership at two million, and Ducks Unlimited is the wealthiest, with annual revenues over $340 million and 700,000 members. All these nonprofits are targeted and aggressive in their approaches to conservation, especially when they combine forces.

The other major difference between now and 1962 is the amount of time troubled bird species may have left. Many are approaching fragile territory—or soon will be, according to the *Science* report data. A consensus is emerging that the next decade alone will likely determine whether the downturns can be reversed—or halted, at least, in the more difficult cases. When Ian Owens, who took over as executive director of the Cornell Lab in 2021, discusses the urgency he feels, he lobs the question back to us: "How long do you think we've got?" We share what we've heard in most interviews, that progress must come within five years, and the next decade will be critical. "Yeah, that's about right," he says. Elizabeth Gray, chief executive officer at the Audubon Society, sees it the same way. "I would say we have a decade to get this right," she estimates. "We have to have a significant turnaround, or I think we will be past the point of no return."

The Bald Eagle is North America's great conservation success story, more than quadru-pling its population in the past decade under an all-out effort to protect the iconic raptor.

3

ERA OF DISCOVERY

Melbourne, Florida

We were just weeks into the first leg of our travels when we stepped out of our trailer on a bright, utterly clear morning on Florida's northeast coast and immediately spotted four Bald Eagles soaring overhead. These majestic birds are unique to North America, and no matter how many times you come upon one, it's a captivating sight. But a small armada, their white heads shining in the sun, is guaranteed to draw a crowd. As we craned our necks upward alongside fellow campers, they started telling stories about all the eagles they'd seen—the nest within walking distance, the pair living on a lake just to the north.

This was a conversation we would have continually over the coming months on back roads across the continent. People love to talk about these magnificent birds, our national symbol, adopted by the Second Continental Congress on June 20, 1782. Our own favorite experience happened to fall on a Fourth of July weekend when we stopped for a few days in Yellowstone National Park at the halfway point in our coast-to-coast journey. Near the center of the park we ran into a traffic jam—a standstill that usually signals a nearby herd of bison or a lumbering black bear. This time, though, neither was in sight. When we pulled off and followed the direction of the stares and cameras, we spotted the attraction: A lone Bald Eagle sat perched at the peak of a lodgepole pine a quarter mile away as if adorning a courthouse flagpole.

So many eagle sightings are no coincidence. The Bald Eagle is in the midst of a resurgence that's spread to every part of North America and has turned out to be the bird conservation success story of our time. In the mid-1900s, the widespread use of pesticides and runaway hunting cut the total population in the United States to just 487 pairs, a number that threatened the future of the species. The eagle began a slow recovery after the ban on DDT, which left their eggshells too fragile to hatch. The Bald Eagle was one of the first birds protected under the Endangered Species Act in 1973. Over the next three decades, the population swelled to 11,000 pairs, and they were taken off the endangered list on June 28, 2007—just in time for the July Fourth holiday. A celebration was held at the Jefferson Memorial in Washington, complete with a Native American blessing and a demonstration flight by a trained eagle named Challenger.

But that wasn't the end of the story—or even its most intriguing part.

In the years since, an aggressive campaign of research and conservation has accompanied the eagle every step of the way. Many eagles today are outfitted with electronic trackers, and biologists follow their travels and monitor their safety. The Fish and Wildlife Service dispatches airplanes to patrol their nesting grounds, and scientists study the birds' DNA to trace their lineage. When it came time to compile the latest eagle count, Fish and Wildlife found a new method. Biologists blended aerial photos with sightings from hundreds of thousands of birdwatchers around the country who are part of a citizen science effort called eBird. The final tally, announced in 2021, far exceeded what anybody expected—316,700 eagles in the lower forty-eight states, an astounding 650 times the number at their lowest ebb. With Canada and Alaska included, the population surpasses a half million birds.

The Bald Eagle's success reflects an era of discovery unfolding in the world of birds at the same time that many species are experiencing dramatic declines. While the eagle's recovery began with the banning of DDT and hunting, its full resurgence came alongside new techniques and technologies that provide a breadth and precision of information unheard of a few years ago. Advances in the fields of genomics, satellite imagery, artificial intelligence, and nanotechnology are reshaping the study of birds. In the past decade in particular, powerful tools have emerged for tracking birds, analyzing populations, and digging into the details of their daily routines. The goal is to inform and ultimately improve day-to-day conservation work on the ground.

The projects are spread around the hemisphere: Fifteen hundred radio towers spanning North and South America ping every time birds equipped with electronic tags pass within ten miles. Two hundred and forty miles above the earth, a satellite antenna attached to the International Space Station can follow birds wherever they go in the world. Advances in DNA research allow scientists to peer into birds' genetic blueprints and decipher their origins and travel histories. A lab in Colorado Springs collects DNA from feathers throughout the hemisphere to determine the routes of one hundred key migratory species. In Hawaiʻi, miniature monitors that work like supermarket checkout scanners clock the comings and goings of seabirds so elusive that until recently they were impossible to study.

This evolution of our relationship with birds—and the laws and policies that developed alongside it—stretches back over a century of twists and turns. As with the Bald Eagle, every major step forward typically follows significant population losses that force rescue efforts for birds heading toward extinction. If there's a single pattern that shows up time and time again in the history of our relationship with birds, it is this: We wait until a crisis to act. And that's where we still are today. There's no downplaying the role of human exploitation as the root cause when birds disappear. And there's no overstating the human ingenuity necessary to bringing them back. Debate and conflict, including at least one homicide, punctuated the decades of progress in this era of discovery. The best place to begin to retrace the story is with the murder.

A Shot Heard 'Round the World of Fashion

From the day he was born, Guy Bradley lived near the center of North America's first massive harvest of wild birds. The year was 1870, a time when a fashion craze for elaborate plumes on women's hats led to a gruesome trade in bird feathers. By the time Bradley turned fifteen, the best feathers, sold by the ounce, were double the price of gold. Sheer greed nearly wiped out stately wading birds like egrets, herons, and spoonbills. As a teenager, Bradley got a job as a guide for plume hunters in South Florida, and on one two-week trip crews killed 1,397 birds of thirty-six different species. Hunters often shot them in their nests, plucked the birds clean, and left the carcasses to rot. When Congress cracked down with the Lacey Act of 1900, which outlawed commercial trade in wildlife that had been taken illegally, Bradley switched sides. In a letter to the Florida Audubon Society, he wrote: "I used to hunt plumes, but since the game laws were passed, I have not killed a plume bird. It's a cruel and hard calling, notwithstanding being unlawful."

The letter, along with his hunting experience and knowledge of Florida's isolated landscapes, helped Bradley land a job as one of the country's first game wardens. He turned out to be a brave and brazen enforcer, single-handedly upholding a hunting ban. He made arrests around the state, sometimes getting into gunfights in the process. Then in 1905, at the age of thirty-five, Bradley noticed a suspicious boat near Oyster Key in the Florida Bay and went to investigate. The next day, a search party found Bradley shot to death, floating in his skiff. The killing was traced to a local plume hunter who admitted his role, but he claimed self-defense, saying Bradley fired first. Even though the evidence showed Bradley's gun remained fully loaded, the suspect was released anyway and the killing never prosecuted. When news of Bradley's case spread around the country, the public was outraged. Meanwhile, as the turn of the century approached, women conservationists boycotted the hats, helped start Audubon Societies in more than a dozen states, and put pressure on hat manufacturers and on Congress. The Migratory Bird Treaty Act of 1918—which made it "unlawful to pursue, hunt, take, capture, kill, possess, sell, purchase, barter, import, export, or transport any migratory bird"—finally banished feathered hats. Nonetheless, the rise and fall of bird populations in North America was just beginning.

The early 1930s found the country struggling with guns and ducks. Unregulated hunting for food and the drought of the Dust Bowl years nearly wiped out the historic abundance of waterfowl. This touched a nerve among the legions of Americans who'd grown up hunting ducks for food and also with a segment of wealthy and politically connected businessmen who'd turned the pursuit into a sport for the privileged. Their influence helped push through a series of actions in Congress that protected ducks and created permanent funding for waterfowl conservation. In the years since, financial support for waterfowl has totaled more than $14 billion and helped pay for the purchase of six million acres of wetlands for the country's 567 National Wildlife Refuges. The money fueled the recovery of duck populations that has made waterfowl the nation's longest running and most significant conservation achievement. Refuge officials and the hunting lobby have jealously guarded those billions ever since.

The nation's postwar boom would soon have consequences for many other birds, including such high-profile species as Brown Pelicans, Peregrine Falcons, Ospreys, California Condors, and the Whooping Crane. As these species began to falter, the federal government ever so slowly began to expand its focus beyond ducks. Most of the research at the time was conducted at the U.S. Geological Survey's research station at the Patuxent Research Refuge near Laurel, Maryland, which would eventually grow into the seat of federal studies of birds. In the early 1960s, a birdwatcher from the Midwest sent the Geological Survey a letter asking whether it was true that American Robins were in trouble. The woman's query was forwarded to a federal biologist at Patuxent named Chan Robbins. He didn't know the answer but decided to find out.

Robbins came up with an ambitious plan to build a databank of bird population counts to see if species were dwindling. To do it, he'd need a small army of volunteers skilled at identifying birds by their songs. These birders would fan out around the country at the height of the spring mating season when the birds are the most vocal—and listen.

Danny Bystrak, a lifelong biologist at Patuxent, was just seventeen when Robbins invited him to participate in one of the earliest surveys in the mid-1960s. Bystrak says Robbins established a strict protocol for "point counts" that is still in place today: Each birder starts thirty minutes before sunrise and covers a roadside route 24.5 miles long, marked with stops every half mile. The birders get out of their cars for three minutes, and write down all the birds they can see or hear within a quarter mile. "You're doing like fifty stops, and basically you're calling on everything you know about birds for four hours straight," says Bystrak, who surveyed the same route for fifty years. He describes the variability among the songs as almost impossible to keep up with. "The deeper you get into it, the more you realize it's worse than you thought—you know, when a Yellow Warbler decides it's gonna sing like an American Redstart."

Within a few years, Robbins's project got a formal name—the North American Breeding Bird Survey—and thousands of the most expert birders wanted to participate. John Sauer, a biologist with the Geological Survey who now oversees the analysis of all the survey data collected, says Robbins's relentless drive kept the project on course. "This guy was the dynamo of twentieth-century ornithology," he says.

The Breeding Bird Survey results became the only baseline for comparing losses among individual species year to year. When the Department of the Interior called for a comprehensive report on the population status of birds in 2007, the survey's data served as the fulcrum for the first "State of the Birds" report the following year. As researchers combined data from multiple sources for the Three Billion Bird report twelve years later, the Breeding Bird Survey remained that starting point.

"Suddenly, you saw the real power of fifty years of data," says Bystrak, "because [the report] wouldn't have been possible without it." Smith, the scientist who did the report's final calculations, says working on the Three Billion Bird project gave him a new appreciation for these archives of population data. "Without those volunteers who go out and participate in the Breeding Bird Survey, we wouldn't know anything about any of this. We're standing on the shoulders of all those people getting up early in the morning, and driving out onto some back roads somewhere, and standing there and counting birds, and writing it on a piece of paper and sending it to Chan Robbins. It's just incredible when you start to realize the effort that's gone into this."

As Robbins was starting the census, another headstrong player emerged who slowly revolutionized the future of bird conservation and rescue. An electrical engineer from Indiana named Bill Cochran was the first to invent a transmitter small enough to ride along with birds in flight. His invention was far from perfect. In order to follow the signals from the transmitters, Cochran had to chase after the birds in what sometimes led to wild pursuits. On one occasion, he was suspected of being a drug-runner. On another, he survived a serious plane crash.

It would take more than a decade for Cochran's idea to become relevant. Only after cell towers and satellite technology matured in the 1980s and 1990s could scientists provide more precise evidence that birds were disappearing. Researchers grew alarmed and have been worried ever since. Unlike Bald Eagle losses, which were caused by specific, known factors, the new culprits were harder to pinpoint.

"We are dealing with a different beast now," says Pete Marra, the bird scientist who helped write the *Science* report. "The problems of conservation now are broadscale, interacting, and insidious in so many ways."

A Shift to Saving Birds Closer to Home

A photo from forty years ago shows John Fitzpatrick in the mountains of Peru on one of his many trips to the Andes. With a tripod balanced on his shoulder, his trademark mustache and longish hair blowing in the wind, he looks like a character sent from central casting—the swashbuckling scientist in the South American rainforest. In those days he spent much of his time traveling abroad, studying birds that had yet to be documented, discovering species for the first time. Fitzpatrick, who would go on to become the country's preeminent bird scientist, says the tools of the profession were basic back then. "Binoculars, a notebook, all done by hand. Computers, we didn't even know the word. A lot of it was on foot. So, our radius of work was about as far as you could walk in one morning."

But change was coming for both Fitzpatrick and his profession. When his first child was born, he decided he needed to stay closer to home. Fitz, as everyone calls him, took the director post at Archbold Biological Station in Central Florida in 1988. That's where he began to realize that the focus of ornithology—unraveling the mysteries of birds, gathering knowledge for knowledge's sake—was no longer enough for the problems he saw birds confronting. Years earlier while at Archbold for an internship, Fitzpatrick had gotten a taste of Florida and had fallen for the Florida Scrub-Jay, a charismatic bird he'd continue to research and protect for the rest of his career. "During this whole time in the Andes, I'm watching Florida get carved up into pieces, watching Florida go to hell basically," he says. "I'm watching the Florida Scrub-Jay dive towards extinction. And I definitely did have a kind of a personal realization that if conservation is going to work, for God's sake, we better be able to do it in the U.S." Working permanently at Archbold, his emphasis gradually shifted to the birds right on his doorstep and to those on the verge of collapse. Those years marked a turn in his overall focus toward questions of conservation. "Science was being directed at how to stop declines," he says, "how to bring back species from the brink."

That shift demanded new research avenues that didn't exist when a pair of binoculars was all you needed. From then on Fitzpatrick, a Harvard-educated scientist with a talent for making the case for conservation, emerged as a force in modernizing his field. After six years at Archbold, he took over as director of the Cornell Lab, which sets the pace for much of the discipline. He grew the lab of about forty scientists, staff, and postdocs to a fully digital operation of 250, added a genomics section, worked with the National Audubon Society to launch the largest citizen science project in the world, and built a wing of the lab devoted to using sound to research birds. This required broadening the staffing expertise.

"We needed all the sciences involved in conservation biology, which is two dozen of them," he says, including more emphasis on bird ecology, biodiversity, mapping, and data expertise. Fitzpatrick stepped down from his lab post in 2021 to become director emeritus. He's the best person to talk about the progress of research and conservation over the last twenty-five years.

"Vastly More Complicated than Rocket Science"

Fitzpatrick's hair is short now, his mustache white, but he's still hiking the Andes Mountains, these days with his daughter, a bird biologist who's followed in his footsteps. If anything, he's even more passionate about the plight of birds today. He's encouraged by the sophistication and vast improvements in research. "That's one of the reasons for optimism," he says. "The tools we have now for understanding birds are way better than they were ten years ago, and they're getting better all the time." The shift from merely understanding birds to actually saving them demands solutions that aren't yet entirely within reach. Fitzpatrick levels his gaze and gets right to the point. "It's not rocket science," he says. "It's vastly more complicated than rocket science."

This has become apparent the more he has watched what works in conservation and what doesn't. "Every individual species has its own ecological story. So, there isn't going to be one solution that's going to fix all birds because they're all different." He uses the example of Bald Eagles and Great Blue Herons, two large birds that have three essential things in common. Both depend on water, feed mostly on fish, and often share the same territories. But for reasons not yet known, their populations are headed in opposite directions.

"What would seem to have been good for the eagle you'd think would be good for the heron, which is rapidly declining across its range," he says. "What's going on there? You can start with the fact that we're not going to find a master solution." Instead, he sees the need for a stepped-up period of aggressive research and conservation. "We've got to get an understanding of what's causing the death rates to exceed the birth rates. It's fundamental because of one thing: More birds are dying than are being born."

Fitzpatrick brings us back to the story of the Bald Eagle to make a point: For all the facts known about high-profile birds, too many species remain shrouded in mystery. The eagle's story is encouraging but it's not a universal model for bird conservation. For starters, the Bald Eagle has a star power and public support that other birds cannot match. Every year since the first eagles built a nest at the U.S. National Arboretum in 2015, the pair has made headlines in *The Washington Post*. The arboretum staff put up a camera to stream footage, allowing the public to follow the action as "Mr. President" and "Lotus" (Lady of the United States) started their new family. Eagle cams peer into nests all over the country, providing millions of people with mesmerizing footage. "I can write letters and tell people how cool birds are," said Charles Eldermire, bird cam projects leader at the Cornell Lab, "but the birds are so much better at it. People only conserve what they know and care about."

Herein lies a fundamental problem: As Eldermire pointed out, how can people care about something they never see or even know exists in the first place? So many beautiful birds survive by hiding, their colors and patterns perfectly matching their surroundings. Others stick to leafy treetops and thick bushes, emerge for an hour or so at dawn and are mostly silent except for a few weeks in the spring. Finding these birds requires patience, resources, and in many cases, sheer luck. But if recovery missions are limited to the favorite high-profile and most visible species, we will certainly lose many more.

We also stand to diminish biological diversity, endangering the engine of the natural world. The heroes of bird conservation hope public perspectives and political priorities will shift, but there's no time to waste. Be they biologists, researchers, summer field interns, or volunteers, all realize the complexity protection requires. Given enough time, money, and advances in technology, top experts in the field say nearly all species can be saved. And at the heart of it all is a basic fact: In order to understand how to preserve birds, first you have to follow them around. This is where the technological advances have been most impressive and where much of the work begins.

Scientists hope to save the Florida Scrub-Jay, the state's lone native bird, with the help of a new tracking technique that follows the birds wherever they go around the clock.

4

FOLLOWING THE BIRDS

Lake Placid, Florida

Nothing moves the first few minutes after dawn among the stunted oaks, skinny pines, and palmetto plants in the sandy scrublands of Central Florida. Then Reed Bowman climbs out of his truck, raises his hand, and starts pishing for birds. He belts out the trusty universal bird call, a simple, steady "shush, shushh, shhhuusshh" of blowing air as if he's trying to quiet someone. It's a lot harder than it sounds. But within minutes, Bowman draws his first families of Florida Scrub-Jays that live in a part of Florida that looks nothing like the tourist postcards. The jays love the desolate, arid landscape stretching over what were once ancient beaches and dunes, dating back over a million years. They fly in at top speed from as far away as a quarter mile, perch in the bushes nearby, watch us, and wait. Forget binoculars. On this February morning, we're all getting the ultimate bird's-eye view.

Scrub-jays, Florida's only native species, are famously curious and often unafraid of people. It takes a close-up look to appreciate the beauty of these spunky birds. They wear their iridescent light blue feathers like tiny sports coats against their grayish breasts. Unlike their crested Blue Jay cousins, a scrub-jay's head is round and smooth. They're almost always in motion, flicking their long tails, cocking their heads, and jealously patrolling their territories.

Bowman is a regular here, so the birds know to sit on the bush and wait while he pishes away. Finally, he brings out the bait and pitches a few peanuts into a clearing. The jays instantly shoot forward to peck at their favorite treats. This gives Bowman just the time he needs to size up his flock as they hop on the ground, turning this way and that, allowing him to spot the colored bands around their legs that identify each bird. "Okay," he says after a few minutes. "That's everyone." This entire family of a half dozen jays is healthy and accounted for at the morning's first stop, so we head back to the truck and on to find the next group.

We joined Reed Bowman and John Fitzpatrick for this monthly census at the Archbold Biological Station to get a look at a creative new tracking technology. Following the flights of birds has become high-tech, and the tools come in different forms to match the bird's weight, range, and the project's scope. One tracker the size of a paper clip fits on migratory birds like a miniature backpack, then clocks a bird's location by tracking the rising and setting sun as it flies along its long-distance route. A network of small towers is spreading to all parts of the hemisphere, picking up signals like tiny air traffic control centers for birds. A satellite system attached to the International Space Station is designed to follow wildlife from anywhere on earth. The common goal is to inform conservation. Nathan Cooper, an ecologist with the Smithsonian Migratory Bird Center, puts it this way: "We can't do anything if we don't know where the birds are."

Florida Scrub-Jays don't migrate, but they do fly around the scrublands constantly. The research team needs to track precisely where they go throughout the day to figure out how to help them survive on steadily eroding territories. The work is headquartered at Archbold, the nonprofit research outpost located near the center of the state. It was established in 1941 by Richard Archbold, an heir to the Standard Oil fortune who loved the natural sciences. Eighty years later, Florida's human population has multiplied tenfold, and 90 percent of the scrubland that jays depend upon has been converted to ranches, homes, businesses, and theme parks. The number of scrub-jays has gone from an estimated historical population of about 40,000 to about 4,000 birds since the 1980s.

For many years, scrublands capable of hosting jays were scattered across swaths of inland Florida in large, contiguous patches. Now more than half of the state's remaining scrub-jays are located in just three core areas near the middle of the state, one right where Archbold is located. Signs of growth are lapping at Archbold's door: Dump trucks hauling dirt to new housing tracts rumble by the front gates every day. Orange groves run up to the edge of its land. All this has made Archbold the perfect place to study the impacts of the state's growth on plants and wildlife in general—and on the Florida Scrub-Jay in particular.

And study they have. Between them, Bowman and Fitzpatrick have devoted eighty years to researching the scrub-jay. As head of Archbold's bird program, Bowman has worked with the toughest bird cases all over Florida. Although Fitzpatrick left Archbold to head up the Cornell Lab, he never stopped researching the scrub-jay. The two have kept detailed notes on every birth, pairing, and death over fifteen generations of jays, first in the office's aging wooden card files before switching to computers. A collection of blood samples traces the relationships of the eighty-five jay families on Archbold they study. They've twice mapped the jay population throughout Florida. But none of this painstaking work solved the problem of how to stop the bird's decline. Both the jays and their scientists were running out of time as Bowman and Fitzpatrick approached retirement. Then four years ago, Bowman came upon a new technology that could follow the jay's every waking move to decipher what might help the birds live on the fragments of remaining scrubland. "This was a huge part of the puzzle we've been studying for all these years," he says.

Now, ten-foot poles are positioned every four hundred yards around Archbold's scrubland, anchored in the sand and topped with electronic receivers. Most of the young jays wear small cellular tags that trigger a signal when the birds pass by. That signal goes to a base station on top of a nearby water tower, and from there into the Archbold computer system. The result is a digital map of precisely how the birds live and move around the clock.

On the day the grid went online, Bowman's research assistant, Young Ha Suh, was testing out this new network of transmitters. The two downloaded the first batch of data collected from the jays' travels over the past few weeks, then watched in awe as an avalanche of signals poured forth. "I said, 'Oh my God, look at all the data that's coming in,'" Suh says. "I'd never seen anything like this. It was really promising." Bowman, tall, bearded, and cerebral, looking every bit the scientist, remembers the exact moment he realized he might be seeing the progress he'd been hoping for. "This is really cutting edge," he says. "Nobody's ever done anything like this before."

The tracking that Bowman and Suh are doing is aimed at analyzing the birds' use of their territories in ways that could be useful for many kinds of birds. The question they want to answer is how best to help the scrub-jays thrive in these close quarters at a crucial point in the jays' development. By following the birds in the two to four years before they become breeders, they can decipher what helps them find partners, set up territories, and become successful breeders. "We should be able to make recommendations to managers about how to keep the most habitat they possibly can on their land," Bowman says, "how to get the maximum numbers of jays on their property." When you're down to just a few thousand birds, the success of each one counts.

A Love-Hate Relationship with Jays

The first time we came upon the Florida Scrub-Jay on a hike just north of Orlando a few years ago, we immediately understood why this bird is so beloved and yet so hard to protect. We were in a county park on a five-acre tract surrounded by new homes, similar to a lot of places the jays live. These tend to be small, hilly islands of sand, white as sugar. The Florida scrub is dotted with some of North America's oldest plants, many of them on the Endangered Species List and all with captivating names like Florida scrub rockrose, whose bright yellow petals open in the morning and close by early afternoon, and nodding pinweed with dozens of red and green blooms shaped like pinheads. The most obvious plants are palmettos, spindly scrub pine trees, and native oak scrubs that are more like bushes than trees.

When we reached the top of a ridge, three scrub-jays, almost assuredly part of one family, landed in the highest branches in a bush merely a few feet away. We were giddy with excitement at what we assumed would be a brief visit, but the birds lingered. After fifteen minutes, we started to move along the path—and the birds did, too. In a busy spot like this, they may have been looking for peanut handouts. (Once people discovered that peanuts will sometimes prompt the jays to jump onto your head or shoulder, word spread quickly and so did the photos on social media.) Or the birds could have been displaying their outsized curiosity. After more than half an hour, an experience we'll always remember, the birds had enough and flew off.

But the scene itself wasn't an entirely happy one. Throughout Florida, much of the remaining scrub-jay habitat has been cut into such tiny chunks it's almost impossible for younger birds to find mates and even make their way to other sections of scrubland. On average, each pair of scrub-jays needs twenty-five acres to establish a territory sufficient for nesting and food gathering. This means that a small cluster of jays like the one we met can face a genetic bottleneck due to inbreeding. If so, they're likely to die out in a matter of years.

Florida remains full of ambivalence when it comes to its scrub-jays. On the one hand, the only native species in a state full of birds is popular in much of Florida. Researchers have posted signs warning people to be careful in jay habitat. They've moved entire family groups out of the path of development to safer territories. Hundreds of supporters have signed up for a fan club called Jay Watch, organized by Audubon Florida to watch over the jays' nests, keep count of their numbers, and draw attention to the bird. Every few years, a fresh campaign gets started to make the scrub-jay the state bird in place of the more ordinary mockingbird.

But not everybody appreciates the scrub-jay. Since they're listed as threatened under the Endangered Species Act, federal and state laws regulate construction wherever jays are found. Developers and some politicians see the jays as a barrier to the growth that powers Florida's economy, leading to a history of friction over how to balance the needs of people against those of the birds. Scrub-jay supporters say the real reason the Florida legislature has so far refused to make it the state bird is that the additional stature would give the jays more sway over development.

Bowman got a taste of the ambivalence when he went to a public hearing in support of an ordinance to help make room for jays in coastal Brevard County near Orlando. The building had to be evacuated and the hearing canceled because someone pitched a smoke bomb into an elevator. "We were going to protect a whole bunch of scrub-jay habitat," Bowman says. "And everybody's afraid of that. They're like, 'Will I lose my access [to develop my land]?'" The conservation plan was celebrated as a strong local environmental stand, but the county commission majority changed hands soon afterward and the proposal never passed.

Property owners have been known to take matters into their own hands, sometimes driving the birds from the land to clear the way for development. Fitzpatrick tells the story of a battle back when he was director at Archbold. One day he visited a nearby property owner whose land was prime jay territory. Fitzpatrick wanted to impress on him what a treasure he owned in a long strip of sand called Lake Wales Ridge. "I took him to the hilltop of these amazing two hundred acres," Fitzpatrick recalls. "I said, 'This is like the gem of the scrub,' that it had the highest diversity of endangered plants in the entire Lake Wales Ridge area, right there."

A few days later, a coworker rushed into Fitzpatrick's office to say a bulldozer was clearing the land he'd just visited. "I said, 'What?' Then I grabbed my big old Panasonic, you know the old video cameras, and I drove down there." When he reached the plot, the bulldozer was churning through the scrub, crushing everything in its path, in clear violation of federal law with no permit and none of the advance legal notice required to work in a sensitive ecology. Fitzpatrick says he was particularly bewildered since the owner had seemed to welcome his earlier message.

Fitzpatrick ran right into the path of the bulldozer. "So, I turned on that camera, and I'm videotaping the scrub-jays right there in the bush, and I start videotaping the bulldozer coming down." The driver came within twenty feet, close enough for a staring contest. "He looks at me, and he turns around and goes back up the hill."

Once the bulldozer driver left the scene, Fitzpatrick raced back to his office and called the property owner. "His secretary said, 'Well, he's in a meeting.' And I said, 'You've got to get him out of that meeting because he's going to want to talk to me.'" When he came on the line, Fitzpatrick didn't hold back. "I can give your company plenty of bad publicity if that's the kind of publicity you want," he said. "After that conversation we had, you're actually destroying that exact piece of scrub?" No further bulldozing occurred, and the owner eventually sold the land to the state.

Some portions of the remaining scrub-jay habitat in Florida are protected as part of parks, conservation areas, or national forests. There are also still many patches of private scrubland that could become valuable habitat if the birds were able to move between them. Passage of the Florida Wildlife Corridor Act in 2021 helped the cause, putting $300 million toward connecting open lands in Florida. That in turn has made the work at Archbold on how to manage these scrublands all the more important.

A Bird's-Eye View

Young Ha Suh's computer screen looks like it belongs on the wall of a modern art museum. Blue, green, purple, and yellow jagged lines fill the screen in an indecipherable design that to Suh, the research assistant working with Bowman, looks like a breakthrough. For two years, she's been gathering the signals collected from the scrub-jays flying around Archbold and working to synthesize the masses of data. "The files are so big it can take an hour just to open a folder," she says as she explains how the data can translate into specific conservation actions.

Suh is part of a new generation drawn to bird research. She grew up in South Korea in an urban neighborhood without much wildlife, but she was captivated by documentaries and family trips to the zoo. She came to the United States to study and is finishing a PhD in evolutionary biology from Cornell University. A tattoo of a Eurasian Magpie stretches down her left arm. It's a favorite bird from her hometown and helps her explain her fascination with birds and technology.

"I've always been, like, what is the bird seeing? How can we understand what decisions make sense to them, not what makes sense to us? This," she says of the jagged map on her screen, "is a proxy for seeing the bird's-eye view of the world."

By tracking the birds as they move about, Bowman and Suh say they've been able to figure out how their birds think. For starters, the jays are highly social. "We found they're a lot like us," Bowman says, explaining that, like many teenagers, young birds spend most of their time in small cliques. Scrub-jays mate for life but as the overall population dwindles, finding partners can be messy. Females are highly competitive and have even been known to play the role of homewrecker, breaking up an established couple for a spot in a comfortable nest. And like patriarchal families, the oldest males will sometimes inherit the best territories from their fathers when they pass on.

The research has also dissected the critical role of fire in the birds' lives. Scientists have long known that fires revive the brush and plants by clearing out overgrown areas and prompting new growth. Jays are dependent on regular fire to provide the low and open habitat they prefer. The birds thrive when fires are set in a piecemeal way so there's a variety of ages to the plants. "So that tells us a lot about how to manage the land," Bowman says. For example, the maps show jays use only portions of the open scrubland, much of which has been altered by the development that surrounds it. The birds avoid fields that are too overgrown or filled with invasive plants. That leads to a fundamental conclusion: Many species, scrub-jays among them, are unable to adapt fast enough and need environments just like the ones they evolved in over thousands of years. As Fitzpatrick puts it: "We're studying how to mimic nature in the best possible way."

The scrub-jay's future will remain fragile in Florida, a state so dependent on growth. But the years of struggle have given this native bird a profile that helps remind people what's at stake with the loss of open land. The work to save birds like the jays plays a part in the push to create a healthier environment for both people and wildlife. After all their years of study, Bowman and Fitzpatrick are both optimistic. "We know everything we need to know right now to save the scrub-jay," says Fitzpatrick, sitting on a picnic table outside the Archbold station the day before the census.

Bowman realizes that not all jay clusters will thrive, particularly on the smallest tracts. But the population overall should be able to remain stable and begin to grow—another species rescued from the threat of extinction. "I absolutely recognize that there are places we're going to lose, and there's not a whole lot we can do about it," Bowman says. "But I also recognize that we have the knowledge and the tools to protect the jays in perpetuity."

Millions of Birds Get Some Jewelry

Laurel, Maryland

The best place to learn about the more global side of bird tracking is the Bird Banding Lab, located at the Patuxent Research Refuge just outside Washington, D.C. Banding is the earliest and simplest tracking form, dating in the United States back to John James Audubon, America's original ornithologist. In the spring of 1804, Audubon described how he first attached thin silver wires to the legs of Eastern Phoebes to see if the same birds would return to his Pennsylvania farm in the spring. Astonishingly, they did. By the early 1900s, researchers realized that banding was a good way to follow birds and understand how their travels fit into the birds' annual cycles.

Today, the Bird Banding Lab, a part of the U.S. Geological Survey's Eastern Ecological Science Center at Patuxent, has turned this age-old practice into tech-driven analysis to map the migratory birds they track on their travels throughout the country. Staffers clamp tiny aluminum bands to the legs of about a million birds a year, then analyze the reports every time one of the tagged birds is recaptured. The staff also monitors all the different tracking tools that are required by law to register with the federal lab. That lets them keep up with every advance in the evolution of tracking birds. "We're in a renaissance of ornithology because of all these new technologies that let us track birds twenty-four-seven," says Marra, the longtime bird scientist with Georgetown University.

We met up with lab director Antonio Celis-Murillo on a September morning during a busy stretch of catching, banding, and releasing birds. The work is tedious and repetitive, but you wouldn't know that from watching Celis-Murillo go through steps he's performed thousands of times for decades. Our visit happened to coincide with the height of the fall banding season when more than 350 of North America's species are migrating south for the winter. Each day Celis-Murillo and a dozen staffers add hundreds of birds to a registry that maps their routes. The surrounding fields of shrubby trees are lined with long rows of nets that look like badminton courts. As goldfinches, kinglets, towhees, and others swarm through looking for food, dozens get tangled in the nearly invisible nylon webs. The banders free the birds and slip them into tiny cloth pouches. That begins a race against a thirty-minute timer that dictates what happens next. "We have to move fast so we don't traumatize them," Celis-Murillo says.

With bags of birds in hand, the staffers hurry to the banding station—a covered wooden picnic table where they weigh and measure the birds and record notes on their overall health. Each bird is fitted with one of the aluminum bands—what the banders like to call jewelry—that is engraved with a unique number that IDs them and is clipped on their legs with a pliers-like tool. Within minutes the birds are registered with the banding lab's database and are on their way. In this round, Celis-Murillo has ended up with an Eastern Towhee. He slowly opens his hands until the bird realizes it's free and flies away. "There's nothing like having a bird in your hands," he says.

In the early years, the lab's work had limited value. There weren't nearly enough banded birds, and for the bands to mean anything, somebody had to recapture the birds and report back to the lab. It was a lot like sending off a message in a bottle. Now more than 6,000 licensed specialists help band birds in all parts of North America, and the project has reached critical mass. Currently 83 million birds have been fitted with the numbered leg tags in the United States and Canada. Bird banding serves as a foundation for tracking all sorts of migratory species. While tracking devices attached to birds can send back data from migratory flights in many cases, those tools provide data only as long as batteries or solar power keep them working. That can be for much of the year, but a percentage of these complex electronic devices will fall off or stop working. The simple bands, on the other hand, will stick with a bird for life.

While quickly clamping an aluminum band around the leg of a Ruby-crowned Kinglet, Celis-Murillo explains how these marked birds help build the lab's population maps. One of the smallest songbirds in North America weighing less than a third of an ounce, the greenish gray kinglet lets you know when he's upset: Bright red feathers that are normally hidden stick straight up on the top of his head. Celis-Murillo works quickly and releases the kinglet, scarlet toupee glowing, before resuming his tutorial on how banding works.

"So there's a chance this bird is re-sighted, and then we have two points for this bird," he says, referring to the kinglet's initial banding and any follow-up capture. "If we look in the database, we may have 100,000 points on kinglets. When you put together all that information and you map and visualize those points, then you start really understanding the patterns."

The deeper analysis is possible when the staff looks back over decades to see where the birds are found, where they've traveled, and what that says about trends in population and migration. "We know birds are biological indicators," Celis-Murillo says. "So if we keep track of their populations over time, we can understand what the problem is now, and we can also look at the past and see how the problem was advancing. But most importantly, with the work of statisticians and biostatisticians, we can project where the future is going."

Whooping Cranes: "It Can Be a Soap Opera"

Gueydan, Louisiana

Sara Zimorski does a good imitation of a Whooping Crane. First, she pulls on a white smock to mimick the crane's fluffy, bleached-white plumage. Her right hand goes into the sleeve of a hand-carved puppet head that's painted to look exactly like an adult, complete with a life-size beak and the crimson patch on its forehead. Then, she slips into the enclosure where the young birds are getting acclimated before they're released into the wild. Zimorski has been at this for so long that when she moves with slow-motion grace among the birds in her crane costume, she can fool them into thinking she's part of the flock. That's when she can fill up their food trays, see if they're healthy, and double-check the three-inch-long tracking devices on their legs that record wherever they go.

Zimorski, a biologist with the Louisiana Department of Wildlife and Fisheries, has been test-driving each successive generation of trackers since her first job with Whooping Cranes twenty years ago. The tallest birds on the continent at five feet, the cranes were among the first species large enough to use the early, hefty devices. Today, Zimorski's cranes are connected to her through low-altitude satellites and the same cell towers used for smartphones. Because these birds are free-ranging and the habitat can be isolated, tracking has been vital to rebuilding Louisiana's crane population over the past decade.

The recipe for the crane's demise is a familiar one—conversion of wetlands and grasslands to agricultural fields, coupled with a century of unrestrained hunting. Only twenty-one Whooping Cranes were living in the entire United States by the early 1940s, six of them year-round residents of Louisiana. State and federal wildlife managers have been working to build the overall North American population ever since. By January 2022, the program reached a milestone of eight hundred birds in North America. The largest crane colony today spends the winter in the Aransas National Wildlife Refuge on the Texas coast before migrating to breed in northern Canada.

In one high-profile experiment, the U.S. Fish and Wildlife Service used an ultra-light plane to lead young birds from a captive breeding center in Wisconsin to a new cluster of cranes in Florida. Pilots befriended the birds, dressing up in the same type of costume Zimorski uses. They climbed aboard planes made to look vaguely like a crane, took off, and guided the birds in the same way a parent would have taught them the route. Dozens of young cranes made the journey over fifteen years, at a cost of about $20 million, but the tactic was eventually halted because the ploy made the birds reliant on humans. The Florida cranes have struggled with breeding and remain a fragile branch of the family, pinning more hopes on Zimorski's cranes in Louisiana.

We find Zimorski at her offices overlooking the 70,000-acre White Lake Wetlands Conservation Area two hundred miles west of New Orleans. The April breeding season has just started, and Zimorski is watching as the first hatchlings begin to stand and start walking. As soon as they're old enough for leg bands, she'll log each bird in *The Whooping Crane Studbook*, a registry similar to that of racehorses, which wildlife managers use to help ensure genetically diverse populations when matching breeding pairs. It's not hard to see why the cranes might thrive in the waterways surrounding the base of the Mississippi River. Wetlands for foraging and nesting extend in every direction. But it's also easy to see the hazards. Paddle-powered crawfish boats troll just yards from the cranes and their chicks, highway traffic is heavy, and farmers navigate tractors and plows along the back roads.

Zimorski had to find ways to keep up with the birds she was releasing and, most importantly, the pairs that settled down to breed. The Louisiana cranes don't roam long distances, but neither do they sit still. The younger birds travel throughout the half dozen nearby parishes, occasionally straying as far as Texas and Oklahoma. The solution for keeping up with them now sits on Zimorski's windowsill: a dozen colorful plastic trackers an inch or so in diameter that look like tiny toys lined up in a row. "The data helps us figure out why some birds are successful and others aren't," she says. "And how can we change things to make them more successful?"

The first thing Zimorski does every morning is open her laptop. Up pops a map with little stars, one for each of her cranes. She clicks on the star of a young bird identified as LW2-20, meaning it was born in Louisiana in the wild, the second chick of 2020. The bird's movements over the past couple of days show up as a dotted line on the map. "So, it's done two exploratory trips, one down there a little west of us, and then just last week, it did this trip over into Texas. Now it's hanging out over here."

Zimorski can tell from afar which of her birds are incubating eggs by how much time they spend on the nest. Some sites are equipped with remote cameras for monitoring select nests close-up. She dispatches biplanes to check on nests in remote places that can't be reached any other way. If the trackers show birds behaving oddly, Zimorski knows something's wrong. One pair of birds that was supposed to be nesting weren't spending much time together. Then she noticed the male was paying attention to a different female. Sure enough, he eventually switched partners. Another male courted two different females long past the time the birds should have made a choice. She laughs. "It can be a soap opera."

We climb into Zimorski's state pickup and drive out to find some cranes amid the flat expanse of rice fields, crawfish ponds, canals, and waterways. Along the way she tells us how birds grabbed her attention as a child, and even now in her forties, they haven't let go. A daughter of two teachers who loved the outdoors, Zimorski grew up in Charlottesville, Virginia, where her parents sent her and her sister to a nature camp that many of their friends couldn't stand. "We both loved it." A high school teacher opened her eyes to the appeal of biology, and her path was set. After college Zimorski took a job at the Whooping Crane reintroduction project in Wisconsin and has worked with various crane programs ever since.

Here in Louisiana amid this fragile restoration, she got so swept up that she found herself working hopelessly long days and often rushing out on weekends when issues arose, such as birds losing radio contact or a property owner hosting a nest needing help. "A couple of years ago, it got to the point where I had to take a break," she says. "I would check my email and check my transmitter constantly. I just had to say, 'I've got to let my mind take a break.'" There's plenty to worry about. The trackers don't always work correctly, and at least one crane rejected hers altogether. "She got depressed and wouldn't fly," Zimorski says.

Overall, the results in Louisiana are encouraging. It took five years before a pair of reintroduced cranes hatched a chick in the wild in 2016 that lived long enough to fledge off on its own. Since then, as many as fifteen chicks hatch each spring, though usually only about half of those survive to leave the nest. Louisiana's total is about seventy-five cranes, but Zimorski says there's always a danger of losing even the adult birds. "Some are going to get killed by predators," she says. "Some birds are going to hit power lines. Some are going to get sick. These things are going to happen."

Whooping Cranes are protected under the Endangered Species Act, which makes harming them a federal crime. Since the Louisiana project started in 2011, twelve birds have been shot in the state. "These are people who are out driving around shooting road signs, shooting turtles from bridges, doing all sorts of stuff they shouldn't be doing," Zimorski says. Here, too, the tracking technology is helpful. The wildlife staff can tell almost immediately that a bird has stopped moving and can usually locate the carcass at the scene of the crime.

In one notorious case in 2016, investigators discovered the GPS tracker pitched into a crawfish pond not far from a knife used to dismember one of the cranes. The perpetrator was sentenced in July 2020 for the death of two Whooping Cranes, fined $10,000, and ordered to pay $75,000 in restitution. The judge said jails were overcrowded because of the pandemic, and so he ordered five years' unsupervised probation and 360 hours of community service with the state wildlife department. "Maybe that will make somebody think twice the next time they're riding around," Zimorski says, "because we're going to do everything we can to catch you."

Finally, we reach the scene that Zimorski was excited to show off. On the edge of a crawfish pond filled with shallow water, a pair of cranes make their way along the water's edge. Right behind them comes a freshly fledged chick—or "colt" as they're called for their ability to run within twenty-four hours of hatching. The tiny bird is a blend of yellow and brown fuzz, about a half foot tall or a tenth the size of the parents that tower above and occasionally stop to nudge their distracted youngster along. What is ankle-deep water for the adults is wing-high on the colt, but the young bird seems to gain confidence and move more quickly as we watch. "They learn pretty fast," Zimorski says.

More and more Louisianans are taking pride in the crane's return, and Zimorski thinks the support would be even wider if more people understood the side benefits. The crane's habitat is also the breeding ground for a universe of fish and shellfish, and the miles of marshes act as a buffer during hurricanes and storms. These marshes are the origins of the watery environment of the Gulf and restoring habitat for these ancient species can help to stabilize the wetlands, balance pressures on the environment, and protect an economic force in this part of the country. "Whooping Cranes can be the flagship species for a lot of others that might not get the attention. So to me, they go hand in hand," she says. "It might be harder to get people excited about some little mollusk or snail, but this five-foot-tall bird that jumps around and dances might be more appealing to people."

Chasing Birds Across the Globe

The idea for the latest evolution of wildlife tracking first came to Martin Wikelski two decades ago while sitting on the edge of the Panama Canal. At the time, he was a postdoctoral researcher in his thirties, studying zoology and helping build a 130-foot radio tower to monitor wildlife in a small section of the Panama rainforest. When a group of friends gathered for drinks one night on the water's edge, an engineer joined them. The man listened as they talked about their tracking project before finally jumping in: "You ecologists are stupid," the engineer told Wikelski. "You have this big topic you could address, but you're thinking way too small."

The upbraiding made a big impression. Not long afterward, Wikelski began a twenty-year campaign to create a satellite devoted to monitoring wildlife from space. In 2020 two cosmonauts attached the antenna at the core of his project to the International Space Station, and his satellite began operating. The official name is International Cooperation for Animal Research Using Space. The clunky title shortens to ICARUS, named for the youth in Greek mythology who learned to fly with wax wings, then made the mistake of going too close to the sun.

Wikelski's satellite is one of two relatively new projects taking global approaches to tracking wildlife. The other is a Canadian-based network of 1,500 radio towers called the Motus Wildlife Tracking System, which has now spread to more than thirty countries. The projects take very different approaches: Motus is steadily expanding its towers across Canada, the United States, Latin America, and Europe. The tower stations will pick up any bird or wildlife equipped with a nanotag that passes within ten miles. Wikelski's ICARUS system is designed to do the same thing—except high enough in space for a global reach. He sees ICARUS as "an internet of animals" because it not only tracks wildlife but can detect the health of species and conditions surrounding them.

The most intriguing part of both systems is how they're trying to invite the public to share in the scientific work. "We think this could be a game changer," says Wikelski.

Almost as soon as ICARUS began its first round of transmissions in 2020, people started tuning in, drawn to the wildlife reality show. Followers in Germany were so taken with a flock of vultures, they carried chunks of raw meat, to bait the birds, to where ICARUS showed the vultures should be to see them in person. Several storks under study in France appealed to followers enough that they made a game out of predicting before the scientists where the birds would land next.

"People are really, actively following these animals," says Wikelski. As people become familiar with the animals they follow, Wikelski believes the animals become meaningful in a way that's more likely to affect people than hearing about the startling loss of billions of birds. "The three billion birds missing is a number that nobody can deal with," he says. "But if you have a hundred individually known animals that are dying somewhere, and you know why and can do something to prevent it, then that's different."

ICARUS is still in the experimental stage. In 2022, the satellite connection, sponsored in part by the Russian space agency Roscosmos, was cut off after Russia's invasion of Ukraine. The project is scheduled to be reconnected in the fall of 2024. Meanwhile, ICARUS is working to create tracking devices light enough for small birds, bats, and even dragonflies. Despite the complications, Wikelski thinks he's seeing something he hopes will disrupt the insular traditions of wildlife research dominated by scientists. "We want people to send us ideas. What do you think we should know about animals? Especially kids, they have fascinating ideas, as we all know," he says. "I think this is the way we're democratizing the science of migration."

Motus is open to the public, too, but in a completely different way. The towers are installed and run by partners who cover a construction cost that averages about $7,500 per station. Partners tend to be schools, research centers, and wildlife refuges, as well as individuals. The radio telemetry that tracks the wildlife is all on the same frequency, which means the data is open to all partners. "This may be one of the best ways to understand the importance of biodiversity," says Stuart Mackenzie, who runs Motus as director of strategic assets. "The public's role in this can't be overstated."

Motus and ICARUS are part of an expansion in tracking technology the U.S. Geological Survey's Banding Lab is watching with enthusiasm. The ICARUS and Motus systems may eventually work together, relaying data across what could be a worldwide interconnected network. This is part of what the U.S. Geological Survey's Antonio Celis-Murillo sees as the next challenge for the field: how to put together broad systems like these to search their combined data for solutions, migration patterns, and obstacles birds face. Since anything attached to a bird for research purposes must be approved by the Banding Lab, Celis-Murillo has a close-up view of the changing technologies.

These days, his lab is processing an avalanche of permits for new tools as tracking techniques push in new directions. The devices are getting smaller and lighter, enabling the tiniest warblers and hummingbirds to carry them. Researchers are experimenting with cameras and sound recorders small enough to fit on birds. They're also capable of tracking altitudes and heartbeats and thus the health of species as they weather the punishing elements of migration. The expanding variety is a trend that Celis-Murillo is delighted to see. "There's a new technology every day," he says. "It's moving so fast I just can't imagine what's coming next."

The California Spotted Owl is at the center of the world's largest research project that uses sound to determine how to protect this storied species.

LISTENING TO THE BIRDS

Sierra Nevada mountain range, California

In 1917, when German U-boats were wreaking havoc on American ships during the First World War, President Woodrow Wilson directed a group of scientists to find a way to locate the enemy submarines before they could attack. The team, which included renowned inventor Thomas Edison, came up with a strategy in a matter of months: The researchers created a contraption the size of a bulky television set that could be lowered into the water behind U.S. ships to pick up the hum of the U-boats. The creation that came to be called a hydrophone turned out to be a breakthrough in the war.

A century later, the audio technology developed during the First and Second World Wars has morphed into the most consequential technique for researching birds across huge landscapes. It took decades of trial and error to perfect the art of listening to nature, experimenting on wildlife from whales to bats and eventually birds. Dan Saenz, a researcher with the U.S. Forest Service in Texas, was one of the first biologists to try analyzing bird songs twenty years ago. He built his recorder in a metal ammunition carton with electronics he bought at Radio Shack. "It was really primitive," he says of his first recorder. "I would take it home at night and write down the results on a pad."

Once Google and Facebook started sharing the software they created for scanning internet photos in 2015, the field called bioacoustics lurched forward. Recordings of thousands of hours of bird songs could suddenly be analyzed almost instantly in the same way the internet tools sort through online content. "That really unleashed the power of these recordings," says Holger Klinck, director of the K. Lisa Yang Center for Conservation Bioacoustics at the Cornell Lab. No part of the study of birds has advanced as quickly, spread as widely, and shown as much promise in its ability to survey wildlife in the most remote and rugged parts of the globe. On the Big Island in Hawai'i, bioacoustics can warn military pilots when there's a danger of striking the island's threatened Nene geese that can congregate and create hazards along runways. At the Powdermill Nature Reserve near Pittsburgh, researchers are building a library of the calls birds make as they take off on migration flights, to explore how they're communicating. In Africa, wildlife managers at certain national parks are listening for the sound of gunshots with the help of audio recorders to catch elephant poachers.

Bioacoustics isn't limited to researching birds, but it's especially well suited for the mechanics of bird songs—a full universe of sounds that can signal everything from where birds are to what they're doing. As recorders have become cheaper and more sophisticated, researchers can post them around a bird's territory and collect sounds for weeks and months at a time. Bioacoustics creates what amounts to maps of vast soundscapes, using artificial intelligence that sorts through reams of sounds to convert into graphics. Justin Kitzes, a computational ecologist with the University of Pittsburgh and a leading academic in bioacoustics, says the discipline's precision and potential are unparalleled.

"I'm well aware that you can talk to any scientist from any time period, and they'll say we're right on the cusp of something extraordinary," said Kitzes, who sees the development of bioacoustics as a vital part of the overall advances in research. "But there's a convergence of a lot of things right now that in the next twenty years are going to be very important."

The most significant test of bioacoustics thus far is playing out in the sprawling Sierra Nevada mountains that run along the eastern edge of the state and are home to the California Spotted Owl. Named for feathers with white splotches and weighing just over a pound, it's a mild-mannered, medium-sized owl with an outsized importance. The California subspecies is a close relative of the Northern Spotted Owl, which ignited what came to be called the "timber wars" in the Pacific Northwest a generation ago. To this day, the Northern Spotted Owl remains an example of what can go wrong in wildlife protection. When the California species showed signs of following the same torturous path as its cousin, wildlife managers realized they had a chance to avoid the same mistakes—if they could figure out exactly where the owls were.

What if the Owls Didn't Hoot?

Connor Wood loaded up his black Chevy pickup with his tent, sleeping bag, clothes, shovels, axes, saws, and four plastic tubs packed with two hundred audio recorders. He headed west from Madison, Wisconsin, in the spring of 2017 with an unusual assignment. Two days later, when the full breadth and height of the Sierra Nevada range came into view, he couldn't help wondering if his experiment would work. The Sierra stretches across 25 million acres, reaches elevations of 14,000 feet, and spans half the length of California. Wood's job was to conduct the first formal census of the California Spotted Owl population in the northern part of the Sierra at a crucial juncture for this bird.

Wood was a good match for the job. He's a perfectionist who plans out every detail. Wood says that may stem from his youth as an Eagle

Scout and his years as a runner in high school and college, where he eventually made team captain. "Statistically gifted," says Sarah Sawyer, the wildlife ecologist who oversees the owl project for the U.S. Forest Service. Zach Peery, Wood's mentor at the University of Wisconsin, who has studied these owls for years, went so far as to name one of his kids after him. When the U.S. Forest Service and state of California decided they needed to launch a new research mission on the status of the California Spotted Owl, Wood got the assignment.

He spent months figuring out how to tackle the Sierra. "I was wracking my brains looking at the size of the landscape and the logistics of surveying this much land. It was not going to work," he says. When Wood and his professor talked through their options, they didn't think the standard approaches using tracking tools or rough estimates would be sufficient. That's when Wood heard about bioacoustics, which at the time was just beginning to pick up momentum. "It was kind of a gamble," Wood says. He started with a small test: He ordered a couple of the audio recorders and set them up in the woods near his home in Madison, where he was working on his PhD. Spotted owls communicate using a variety of hoots, whistles, chitters, and barks, which by all accounts Wood has thoroughly mastered.

"I went around and hooted at it from different distances and different times of day," he says. When he scrolled through digital versions of the recordings, he was impressed with how crisp and clear they sounded. "We thought it was not such a crazy thing after all," he says. So in the early spring of the year, Wood put together a field crew, ordered the recorders, and arranged for the team to begin work. He was still apprehensive. "We were wondering how much these owls would actually hoot," he says. "What if it turned out they were really quiet? Then this whole thing would fall apart."

The Sierra Nevada is a blend of breathtaking beauty, formidable peaks and crags with a geology that serves as an engine of California's economy. The mountain snowpack feeds lakes, rivers, and streams, producing 60 percent of the state's developed water supply. That includes the four-hundred-mile Central Valley just west of the mountains, which is the most productive farmland for vegetables, fruits, and nuts in the country. The range holds four national parks, including Yosemite, and twenty national wilderness areas that attract more than 22 million visitors a year. The competing demands put pressure on the Sierra in general and its wildlife in particular, none more than the California Spotted Owl.

These owls live in forests of ponderosa pines, cone-bearing conifers, and white fir trees that can take centuries to mature. Growing as big as eighteen or nineteen inches tall, the owls are nocturnal birds that feed on small mammals, mostly flying squirrels and wood rats. Because of their preference for old-growth trees, the Spotted Owls are barometers of the forest's overall health. As the oldest of these forests have mostly disappeared, so have the owls, losing more than half their population over the past half century or so until just several thousand of the birds are estimated to remain today. "The physical structures of the forest that the owls need are mostly gone," says Wood, "and my great-grandkids would be dead before they're back."

Wood's first weeks in the Sierra were slow-going. It was still early enough that a mix of snow and mud covered the backwoods roads. On one of his first trips in the mountains, he and his team got stuck in a snowbank and had to dig their way out. His plan was to complete an initial assessment of the owl's status by strapping the two hundred recorders the size of lunch boxes around the lower trunks of trees in thousands of acres of the forest. Then they'd move them each week, hopscotching across much of the northern part of the range during the spring and summer when the owls are most vocal. The recorders are timed to pick up hoots and calls from dusk through dawn and store the files on digital cards that are later downloaded for analysis. The tricky part is training the software to identify the owl's unique sounds. Once they'd nailed that, Wood could pinpoint their rough locations.

The Barred Owl is an intruder that has disrupted the fragile population of California Spotted Owls. This is a tale of two owls playing out in the Sierra Nevada mountain range. The Barred Owl has migrated across the continent and now is a troublesome invasive species in California, the Northwest, and Canada.

It didn't take long to piece together a picture of what was happening to the birds. Researchers already knew wildfires, drought that diminished food supplies, and encroaching development were threats. But there was more: By the end of the summer, the recordings made it clear that the California Spotted Owls were facing a potentially lethal opponent in Barred Owls long known to be moving in from the East. As soon as Barred Owls stray this far from their normal range, they become the equivalent of an invasive species. These owls are larger, more aggressive, and can easily displace or even kill a Spotted Owl. Spotted Owls mate for life, and Barred Owls break up pairs and disrupt mating. A Spotted Owl sometimes goes as long as six years between broods, so any disruption is a significant threat to the species' long-term survival. All this gives the spread of Barred Owls the potential to alter the balance of nature in the entire forest, undermining not just the Spotted Owl but an assortment of wildlife. If the trend continued, the California Spotted Owl would soon face an accelerating downward slide that would almost certainly require listing under the Endangered Species Act.

In December 2017, Wood and Peery flew to San Francisco to meet with Sawyer and the top U.S. Forest Service leadership who oversee the fate of the owl. Wood told them the findings were only an initial assessment. In order to come up with a rescue plan, he needed to post recorders in every part of the owl's Sierra range and build a precise map to locate all of the California Spotted Owls and the Barred Owls, too. Wood suspected it would take time to get a response from such a big group, as well as their superiors, on a proposal that could push the annual cost up close to $1 million a year. So he was surprised, as well as a little daunted, when Sawyer told him at the end of the meeting to go ahead with the project right away. Wood would need to create a tenfold expansion of the survey, covering millions of acres of forest and making it the largest bioacoustic undertaking in the world. Sawyer told us that her group realized how much was at stake and how little time they might have. "We thought, if we're going to do something about this, we'd better get on it," she says.

The Timber Wars

It takes Sarah Sawyer a few minutes to settle down when we meet for lunch in downtown Sacramento, not far from where she lives. It's been a hectic morning with her two young boys. "It's a balancing act," she says, which seems to apply to both her role as mom and her job overseeing 22 million acres of public lands from California to Hawai'i. "Okay, we're going to make this lunch last as long as possible," she decides. She's joking, but in fact the full story of the Spotted Owl stretches the length of her lifetime and includes a roiling mix of politics, science, violence, and sabotage that takes time to sort through.

Sawyer began her career abroad as a field ecologist in stops from Haiti to Africa, where she met her husband while both worked in the Congo. When they returned to the United States for Sawyer to complete her doctorate, she brought a conviction with her that forests need to be shared for recreation, timber sales, wildlife protection, and water. "What I learned in Africa is that's how it has to work when people are relying on those resources for survival—day-to-day survival," she says. But it also applies in wealthy countries like the United States where public lands are under pressure from all sides. "Everyone has a different idea of what they want to see out of their national forests, and those opinions are all valid," she says. "But how do we create the science that helps us sift through all those opinions and then use it to manage in the best way?"

Sawyer sees the controversy over the Northern Spotted Owl as a cautionary tale. While the origins date back to when she was a schoolgirl, she's learned enough of the Northern subspecies's story to know she needed to take a very different approach this time around. "Let's use everything that happened over the decades in the Northern Spotted

Owl territory to do it much more rapidly and efficiently in California, so that we don't get to the point where it needs to be listed under the Endangered Species Act, so we don't get to the point where the Barred Owls are everywhere and they're impossible to deal with." She stops for a moment, then adds: "We need to take all those lessons learned during the timber wars."

Three subspecies of Spotted Owl once filled the ancient forests from British Columbia to Mexico. Only minor differences in genetics and feathers separate the Northern, California, and Mexican Spotted Owls. But the Northern was the one that drew all the attention after it collided with the timber industry in Oregon, Washington, and Canada. Logging had long been a mainstay of the Northwest economy, growing up as settlers moved west and expanding with the country's waves of growth. The grueling work of felling trees was considered a triumph. We mythologized Paul Bunyan as a heroic, seven-foot lumberjack who could clear an entire forest with a single sneeze. "As the traditional lumberjacks died off, Paul Bunyan migrated from the forest bunkhouse onto the pages of children's books, where he symbolized American power and masculinity," reads an essay from the Wisconsin Historical Society. The growth marched forward from the Gold Rush, through the western expansion of railroads, to the postwar housing boom. These events propelled the logging industry into taking what was needed from the land during a time when resources seemed infinite. At the height of the forest harvesting, the Audubon Society estimated that loggers were taking enough old-growth lumber each year to fill 20,000 miles of trucks lined bumper to bumper.

Then in the 1970s, as environmental awareness was starting to spread, researchers began to document the impacts of the loss of forests on wildlife. A young biologist with the U.S. Fish and Wildlife Service named Eric Forsman became captivated by the Northern Spotted Owl's struggle. He eventually led a study that recommended the owl be protected under the relatively new Endangered Species Act. At about the same time, a group of environmental activists looking for a way to curb logging in national forests realized the owl's fragile status could be their answer. It took almost a decade, and years of legal battles, but the Fish and Wildlife Service finally listed the Northern owl as threatened under the act in 1990.

The reaction was explosive. The decision pitted the owl against the logging economies central to local communities in the two states and ignited a wave of protests, violence, and sabotage. Environmental activists planted themselves in the path of logging operations, anchored themselves in concrete and sat for weeks up in towering trees slated for harvest. Loggers and their supporters marched in counterprotests and filed a series of lawsuits over the limits on harvesting the trees.

The owl made the cover of *Time* magazine, with the article posing the case as more than a contest between a bird species and an industry. "The owl battle is an epic confrontation between fundamentally different philosophies about the place of man in nature," the *Time* story proclaimed. "Are the forests—and by extension, nature itself—there for man to use and exploit, or are they to be revered and preserved? How much wilderness does America need? How much human discomfort can be justified in the name of conservation?"

The friction reached bizarre levels. Death threats came in against Smokey Bear and Woodsy Owl ahead of the traditional Rose Festival Parade in Portland, Oregon. Every year foresters had donned the costumes to appear in the parade, but faced with the possibility of violence, the service decided it was best that year to leave Smokey and Woodsy at home.

In 1993, President Bill Clinton, Vice President Al Gore, and a third of their cabinet traveled to Portland for a summit to resolve the standoff once and for all. That led to a compromise called the Northwest Forest Plan. It set aside land for the owl's protection in exchange for allowing one billion board feet of lumber annually—about a fifth of the take at its height—to be cut and sold. The controversy gradually settled down, but it has never fully ended, and its impact on the Endangered Species Act is still evident today.

Although the owl clearly deserved protection under the letter of the law, federal agencies struggled to withstand the political pressure against limiting logging. Decisions took years to make, including the many bureaucratic steps associated with listing the owl as threatened. As time went by, and the White House shifted from Republican to Democrat and back, stances on the owl swung back and forth as well. In the most recent instance, President Donald Trump's administration lifted prohibitions on logging on 3.4 million acres of Northern Spotted Owl habitat in national forests in January 2020, a ruling reversed by President Joe Biden the following year.

Neither the region's lumber industry nor the owl have fared well since. Logging never met the goals set under the plan, and thousands of logging jobs were lost. Meanwhile, the arrival of the Barred Owl moving west, first to British Columbia and then down the coast into the Northwest, undermined any chance the Spotted Owl had of recovering in the Pacific Northwest. The Barred Owl is a treasured species in the East, famous for its hoot that sounds like, "Who Cooks for You?" But as it moved west, the owls turned out to be every bit the disruptor experts feared. In Canada, the Spotted Owl has all but disappeared from the wild. The last few birds are in breeding centers as biologists try to rebuild that population. In Washington and Oregon, about 4,000 Northern Spotted Owls remain, the lowest population yet. "It's not looking very good," said Forsman, the wildlife biologist who first studied the owl in the late 1970s. "The populations seem to be gradually going downhill, and it's not clear if or when that's going to stop."

Artificial Intelligence and a Million Hours of Hoots

In the upper reaches of Yosemite National Park, the California Spotted Owl acts as if it has been expecting our arrival. We'd been hiking all day in the Sierra, checking in on the network of audio recorders with Sarah Sawyer and the project's field supervisor, Kevin Kelly, as they sized up the state of the forest. Our owl is right where Kelly's staffers predicted it would be, in a giant hardwood so tall we can't see its top. The owl is perched on the lowest branch, taking us in as its round brown eyes open and close in what looks like boredom.

After weeks of talking about the bird in interviews, it's thrilling to see the creature at the center of this rescue work—and to get a glimpse of its easygoing nature. At one point, the owl suddenly flees its branch and we think our visit has ended, but then it lands a few yards away and continues the staring contest. These owls are known for their friendliness, to the point that some wonder if that hurts their chances of survival. From time to time, the owl raises its powerful talons, scratches a spot or two, and reaches back down to sink its claws into the bark. Eventually we must move on, and the bird's big eyes follow our departure.

Walking in the forest with Sawyer is like visiting a museum with its director. She identifies the trees, rocks, and soil types as we go, but she's actually looking for the deeper story of the woods. In places, the trees are so thick and tall the forest is dark even in the middle of the day. The air is filled with a rich, musty smell of decaying branches and ferns. The forest floor is scattered with tree sprouts trying to find enough sunlight to survive. In one section of the Stanislaus National Forest adjacent to Yosemite, she stops to look over the damage from an old wildfire. She checks the health of the tree bark and branches in the midst of a season of both drought and wildfire. Sawyer is responsible for overseeing the health of millions of acres of national forest, home to countless species of wildlife, but she has singled out the Spotted Owl to experiment with different tactics. "It's safe to say that the traditional approaches haven't been enough to deal with what's going on," she says, adding, "so this is the really big one for me."

Our visit brought home the project's massive breadth. We were able to check on just a few of the thousands of recorders spread throughout roughly five million acres of the Sierra. The recorders are spaced in a way that covers the owl's likely territory throughout the mid-level of the mountain range. The project was in its fourth year at the time, which meant the research had already revealed much about where and how the owls live. The owls would sometimes venture into the territories of competitors to forage for food, and Wood worried whether this would lead to double counting if the same bird were picked up by two recorders. Another team from Wisconsin put mini-recorders on the backs of selected owls and discovered they make almost no noise when they're raiding other tracts. "That was really, really good news," Wood says. They added remote cameras to some of the nests to observe the young and study what and how the parents fed their owlets. All this meant that some of the Spotted Owls were under surveillance three different ways at the same time, helping build a portrait of every angle of their lives.

Half the job is capturing the sounds of the forest; the other half is identifying the owl calls amid what eventually totaled a million hours of recordings each year. Here is where advances in artificial intelligence make all the difference. Algorithms use a digital imprint of the owl's hoots to scan the recordings at ultra-high speeds and flag any sounds that seem to match the owl's calls. Then members of the field team sort through the flags to confirm whether it's actually an owl, which then becomes part of the mapping data. It can be a tedious job. Kelly tells the story of one recorder that happened to be located next to a herd of cattle on forest land. One had a cowbell that clanged almost constantly, filling the recordings with the incessant ringing. The algorithms scan out extraneous noise but didn't know what to do with a cowbell. Kelly, who got the job of monitoring those particular recordings, is used to dealing with odd sounds. "But that was the worst one I've ever had," he says. "Something about the cowbell triggered the detector. I went through three or four thousand cowbell hits."

Before the end of 2018, Sawyer and Wood knew the project team had to make a decision about the fate of the Barred Owls in the Sierra.

There was really only one option to halt the invasion: They'd have to begin killing the Barred Owls to save the Spotted Owls. That's what managers had started doing in the Northwest about a decade ago, and the Fish and Wildlife Service agreed to a test in California. "No one wants to be shooting owls. It's like the worst-case scenario," Wood says. But he and Sawyer conclude that the recordings predict disaster on the horizon. "We can look at the Northern Spotted Owl and see that when the Barred Owls reach a certain density, they just decimate the Spotted Owl population," says Sawyer. "We have no reason to think it would be any different with the California Spotted Owl."

"It's Got to Be Done"

Nothing is left to chance when it comes to killing one owl to save another. Nick Kryshak and Danny Hofstadter, two veteran biologists who got the assignment, spent weeks in training. They learned to fire from a precise distance—close enough to kill the birds instantly, not so close as to do significant damage to their plumage. They use a light gauge shot so the owls can be studied and perhaps used in taxidermied form in a museum afterward. But there's no getting around how strange it feels to shoot a species that on the other side of the country is considered a cherished bird. Kryshak and Hofstadter both were startled the first time they watched their trainer take aim at one of the owls. They were standing not far from where the Barred Owl was perched when the shot rang out. "I just remember the air was suddenly full of feathers," says Kryshak. "Coming down like from a snow globe for the next two minutes." Hofstadter shakes his head: "It was weird."

The shootings became routine after they'd been at it awhile. The bioacoustic mapping marked the territories where the Barred Owls lived. Kryshak and Hofstadter usually located the birds easily, or sometimes by playing a recording of the owl's own hoots to get the bird's attention and prompt a responding hoot that gave away its location. Still, the work sometimes took its toll. In one case, Kryshak had to kill an owl he'd gotten to know from past research visits. It was a large, magnificent creature with a bright yellow beak and huge dark eyes that no doubt could see Kryshak in the dark as he lifted his gun, took aim, and fired. The bird dropped to the ground. He loaded the remains in his cooler and made note of the details. But he'd had enough. "I went home for the night. Called that one a day," says Kryshak, and then adds: "We're damn near positive we're doing the right thing."

The idea of killing Barred Owls draws plenty of criticism. Before the shootings began in the Northwest, the Fish and Wildlife Service overseeing that project consulted with an ethicist to work through the issues. Animal rights groups filed three separate lawsuits to try to stop the shooting. But veteran researchers say we've long since reached the point where tough interventions are necessary. "We manage species using lethal means all the time, whether it's coyotes or ravens or crows," says Zach Peery, the University of Wisconsin professor working with Wood on the research. "This is just the latest iteration of species that have moved in from human activity. Certainly the public has conflicting emotions because they just don't see the division between native and invasive species quite as clearly. They may not understand the magnitude of the biodiversity loss, but everyone in the scientific community knows it's got to be done."

As Kryshak and Hofstadter watched the dueling owls close-up, they also grew convinced. Up to ten times the number of Barred Owls can live in the territory of a single Spotted Owl. They can feed off all kinds of prey, produce many more young, and even mate with Spotted Owls to create a hybrid with the Barred Owl's aggressive style. "If we don't do these removals the Spotted Owl will go extinct," Hofstadter says. "There's no doubt about that at all."

Their work also confirmed that the geography of the Sierra may end up helping to save the California Spotted Owl: In the Sierra Nevada range, the Barred Owl's entry is blocked on three sides—the Central Valley to the west, the high mountains to the east, and the desert to the south. That leaves one route, through the northern rim of the Sierra. Kryshak and Hofstadter found it helped to use an analogy borrowed from the *Game of Thrones* series that features a huge wall on its northern boundary. "We just have to keep them from coming in through the north, because that's where they are," Kryshak says.

It also helps that they caught the invasion early. They shot fifty owls in 2019, ten in 2020, and just four in 2021. By then, almost all the Barred Owls were gone from the Sierra. The recorders documented the rise and fall of the Barred Owl population. In 2017, 8 percent of the recorders picked up Barred Owl hoots. The following year, that number jumped to 20 percent as the invasion accelerated. After three years of shooting the Barred Owls, the number dropped to 3 percent and they halted the killing. "The Barred Owl has been functionally eliminated from the Sierra Nevada," Wood says.

The future of the California Spotted Owl now looks far more secure. There are several scenarios for how the birds will fare depending on everything from wildfires to how much forest management is done. Wood suggests the population declines could continue for a time as conservation work ramps up, and then level off and begin to rebound in coming years. But already, the outlook has improved significantly with the Barred Owls gone. "That really represents a pretty big conservation victory," he says. The most encouraging development has been watching the Spotted Owls move back into territories where they'd been pushed out and pick up where they left off. "When we remove the invasive species, they're back at half the sites within a year," says Sawyer. "That's really encouraging."

The owl project is now widening its focus to find ways of supporting all the major bird species in the Sierra. Wood is adding nearly one hundred new bird species to the bioacoustic analysis. Sawyer, who has since been promoted to national wildlife ecologist for the U.S. Forest Service, hopes to use lessons from the California owl project in the 193 million acres she now oversees. Sawyer and Zach Peery have also won a $1 million contract with the National Aeronautics and Space Administration to combine the agency's satellite imagery with their bioacoustics research. This will enable Wood to map the full landscape of birds against the topography and forest structure of the Sierra. They hope this will give state and federal foresters new ways to manage the land and confront future threats. The most pressing is the rise of wildfires that have done extensive damage in California, including to the owls' territory in the Sierra.

"This is such a special place for so many people and so many organisms and so many things that exist only here," Wood says. "If we can protect this area and learn from what we're doing, we can use these lessons with all kinds of birds in all kinds of places."

Postscript: Our Personal Owl

From Beverly:

Our neighborhood Barred Owl reminds me of a mom calling the kids in for supper. "Who cooks for you?" the bird asks. "Who cooks for you-o-o-o-o?" The owl's dinner bell is reassuring in its consistency, and I sometimes laugh out loud when it catches me off guard.

I take pride in having a personal owl. He makes me feel secure, like all's well in my little patch so long as the wooded buffer surrounding our condo is hospitable habitat for a magnificent owl. Plus, his daily hooting makes me happy. There's menace in it, too—a plaintive melody with an underlying eeriness that's hard to describe and even harder to imitate. But lots of people try. We are touring the cattle ranch that the Archbold Biological Station operates as an experimental laboratory when director Hilary Swain tells us there's one nesting nearby. She stands up, takes a deep breath, and belts it out: "Who cooks for you? Who cooks for y-o-o-o-o?" (It's that elongated second phrase following so quickly behind the initial staccato line that makes it comical.) As we applaud, she seems pleased and takes a little bow.

Barred Owls are tree-trunk brown with a flat face and enough scattered white markings that it can disappear as it perches in the 90-degree angle between a branch and trunk. Our owl lives not fifty yards from the porch and yet we'd never seen him the first months we lived here. Then late one afternoon, I was trying to lure a Baltimore Oriole by playing a recording of its song on my phone. Out of nowhere the owl arrives in a blur, catching my attention with a swoosh as he lands on a nearby oak.

Being careful not to take my eyes off him or make any sudden moves, I speed-dial Anders. He's a few yards away with a camera lens the size of a small telescope. "Come here," I whisper into the phone. "What?" he asks. "Come here," I repeat as loudly as I dare. "It's the owl." He hustles over and clicks photo after photo for the next ten minutes as dark settles in. This lens is too heavy for long hikes, and so it's a rare opportunity to get a granular view. "It was just unbelievable," Anders will tell you. And it was, as you can see from the Barred Owl photo toward the beginning of this chapter.

That evening, thinking about the episode after the euphoria wears off, I realize something. Our Barred Owl arrived at dinnertime, seconds after hearing the Baltimore Oriole's song. If it had been real, that unsuspecting, gorgeously orange-and-black songbird might have been the owl's hors d'oeuvre. Served up by me. This is nature's underbelly, an unsavory side we've bumped up against in differing scenarios. We never get used to it.

In Yosemite, Sarah Sawyer explained how the Barred Owl's westward march was made possible by the wooded corridors humans created that once were plains. By the time it reaches California, the Barred Owl becomes a new predator, clawing into the territories of the cute, defenseless Spotted Owl. It couldn't help but taint my regard for our neighborhood specimen. This owl is not a pet.

The marauding Barred Owl is just one of scores of invasive species, from wiregrass to cats and mosquitoes, that disrupt an ecosystem in any number of ways. While I understand the necessity for shooting the owl, the reality is hard to swallow.

Working on this chapter in the quiet of home, it seemed like every time I got comfortable, our personal Barred Owl started to hoot off schedule. Once I was literally typing the words "had to kill an owl," and here it came, "Who cooks for you-o-o-o-o?" Each time it urged me to decide: Friend or foe? It's tough to sort out.

The Great Blue Heron, shown here with its decorous mating plumage, is one of the species most often reported by birders as part of the massive citizen science project powered by the eBird smartphone app.

6

CANARY IN THE COAL MINE

Rock Creek Park, Washington, D.C.

Scott Stafford catches a glimpse of a bird darting along a stream in Rock Creek Park during his morning walk through one of the best places in Washington to watch the migration parade go by. "There he is," he says. "A Louisiana Waterthrush." The bird with a chocolate brown back, a white slash over its eyes, and a streaky white breast vanishes as quickly as it appeared, flitting farther down the creek bed where it likes to hang out to snag insects. The warbler may be gone, but this encounter isn't over. As we amble along, Stafford thumbs his way through a couple of clicks on his smartphone to transmit his find to an online site called eBird. Within minutes, his waterthrush will be one of hundreds of millions of sightings that fuel the largest citizen science project in the world.

The same technology reshaping bird research is doing double duty for birders like Stafford. The Cornell Lab's bioacoustics software that lets Connor Wood pinpoint the location of California Spotted Owls is also a smartphone app that can tell birders which species they're hearing. The weather radar that helps ecologist Adriaan Dokter track clouds of birds can guide birdwatchers to the heaviest migration days via a free service called BirdCast. The eBird applications that scientists use to count the Bald Eagle also let Stafford track the birds he sees every day around his home.

The most intriguing part isn't how technology is serving birders; it's how these new tools are helping birds. Every time someone sees a hawk and reports it to eBird, for example, they're contributing bits of data to the Cornell Lab. The lab, along with the National Audubon Society, created the system a quarter century ago. Without fully realizing the potential, Cornell and Audubon trained a network of hundreds of thousands of birders to be roaming sensors collecting data around the globe. The extent to which these citizen science contributions can be the basis of full-fledged scientific research is only now becoming clear.

Stafford is the perfect example of how this mutual arrangement works. We join him for an early-morning hike through Rock Creek Park in mid-April 2022. It's still early in the spring migration, and the newly arrived White-throated Sparrow is belting out its thin refrain that sounds like "Oh-sweet-Canada-Canada" or "Old-Sam-Peabody-Peabody." Stafford is out almost every day, especially during the migrations or in early summer to watch as birds nest and forage. A computer programmer in his early forties who works from home in Northwest Washington, Stafford also keeps an eye on what's happening around his yard throughout the day. "So I'm right by the window," he says. "I see doves on the deck. Yesterday there was an American Crow building a nest, and the House Sparrows are collecting nesting material, and the woodpeckers are going into their holes. It's an entire world out there that's absolutely fascinating." He's filed some 3,600 eBird checklists, as these sightings are called, and consistently ranks in the top twenty eBirders in the capital area.

What does Stafford get for his trouble? eBird is like a personal databank and birding guide rolled into one. The app records every bird he reports in a personal running tally called a "life list," including exactly where and when he saw them. eBird is doing this for everyone else, too, and shares the data in lots of ways that help birders like Stafford decide where to go next. He can look up what other eBirders are seeing, check maps of the active "hotspots," and get email notifications when a rare bird is reported nearby. "Every day, I'll be on eBird to see what's coming next," he says. "eBird is how I do everything."

New digital tools for birders continue to grow more sophisticated. One of our favorites is a five-by-four-inch metal box that sits on our porch and lets us know what birds are around even when we're asleep. Called "Haikubox," it records every song or call from birds within tweeting and chirping distance, and via WiFi automatically logs them for us on an app and our computers and sends the data to the Cornell Lab. We can see the list and listen to the stored recordings whenever we want. It also links to every other Haikubox so you can check out what others are hearing. A couple from Sarasota, Florida, David Mann and Amy Donner, both biologists who studied at Cornell, developed the Haikubox with support from the National Science Foundation and now sell the tool on their website. "I think it can engage people in new ways and help people to get into birds," says Mann.

If you just want to know what's flying around your neighborhood, the National Audubon Society released a different sort of free bird tracking site in the fall of 2022. Called the Bird Migration Explorer, Audubon has woven a half dozen different bird counts and data collection projects into an elaborate map that filters its content in all sorts of ways, including by zip code. "We believe that people ultimately care about what is going on in their own backyards," says Chad Wilsey, Audubon's chief scientist. "The initiative meets people where they are with the birds they love, with what's local to them."

Of all these tools, the project that has developed the fastest and furthest so far is eBird, which has spread to two hundred countries and has collected more than a billion bits of data on where and when people see birds. As birders logged more and more data, the concept expanded into a compendium of maps, animations, lists, and databases. eBird is now starting to use all of this information to guide conservation, help wildlife managers oversee species, and provide the raw material on bird populations and migration to research scientists. eBird's connection to birders remains crucial, since their smartphone clicks of sightings are what feed the entire system. But Amanda Rodewald, the Cornell Lab's senior director of the Center for Avian Population Studies who oversees both the forty-member eBird staff and its role in the marketplace, says the lab believes this experiment that started on a hunch in the late 1990s has the precision to produce far more targeted efforts for saving birds. "We now have the opportunity to transform how we do conservation," she says.

"We Had No Idea What We Were Doing"

Frank Gill was the chief scientist at the National Audubon Society in June 1996 when he called up his friend John Fitzpatrick, who'd just taken over as head of the Cornell Lab. "He asked if we shouldn't talk about some kind of internet thing to create some sort of checklist," Fitzpatrick says. Gill laughs about the early days and explains it this way: "We had no idea what we were doing. Neither of us knew anything about computers or technology, but we were spurred on by what was happening with the internet." Within a few months, their staffs had put together the first balky version of a program they called BirdSource that asked birders to post their sightings on this new website.

The pitch to birders was simple: Send us your checklists and you'll be helping birds. But it wasn't an appeal that motivated many people. The project dragged along in its first years with just a few thousand visitors a month, not enough information to accomplish much. In 1999, the Cornell Lab landed a $3 million National Science Foundation grant and took sole control of the project. But it didn't find its way until 2005, when the lab decided to hire people who were both skilled website developers and birders. "They understood what birders wanted and needed," Fitzpatrick says. They changed the pitch of what was now called eBird: Tell us what you see, and we'll organize your birding life. Instead of being a duty, eBird turned into a kind of game that leveraged the desire to collect a life list of bird sightings and the competitive nature of many birders. "That next month it took off," said Chris Wood, eBird's managing director. "It was immediate."

The appeal was lasting, too. eBird has grown by as much as 20 percent each year. More than 800,000 contributors have sent in some 75 million checklists to date, making eBird a global citizen science effort. Even so, birders aren't perfect data providers. People tend to go birding in attractive and accessible places, while on vacation and in parks and preserves. So the lab figured out how to use data from state agencies and satellite images from space to extrapolate what bird populations are likely to be in places where sightings from birders are scant.

Over the past decade, eBird's data and maps have become reliable enough to be used for scientific and academic research. Fitzpatrick showed us eBird data maps that can pinpoint where more than five hundred species of birds are found at any given time in North America. These maps have evolved from broad brushstrokes of population estimates a few years ago to granular, pixel-rich graphics that locate where species and their abundance are down to within a few miles. "I can't begin to describe to you how profoundly important this new tapestry is," says Fitzpatrick. "We've never come close to having this level of elegantly produced, quantitative knowledge about what different populations are doing."

This makes it possible to see where birds are thriving and where they are not, allowing researchers to determine where problems are on the ground. One of the best examples of how eBird is helping to guide conservation is in California's Central Valley. The valley is among the nation's most productive agricultural corridors, but it's also a critical stopover for millions of migrating shorebirds. That put the two roles at odds for years until The Nature Conservancy came up with a Solomon-like solution, one that makes room for birds in the middle of dozens of farms along the valley.

Pop-up Stops for Birds

Colusa, California

On our way to get a closeup look at this intriguing conservation idea, we almost never lost sight of the olive groves, almond trees, rice fields, and vineyards as we traveled the 450 miles from one end of the Central Valley to the other. But long before we reached our destination an hour north of Sacramento, the nature of the dilemma was on display around us. The Central Valley has a unique geology that drops to just above sea level in places, with fertile soil beds built up from centuries of rich sediment flowing off the mountain ranges on either side. A century ago, more than four million acres of marshes, bogs, and swamps covered about a third of the valley that runs up the center of the state. But they were gradually converted to orchards, farms, ranches, homes, and businesses as the valley's population grew to 6.5 million people.

Today about 200,000 of those original acres of wetlands remain, a shortage The Nature Conservancy has tried to address by buying strips of wildlife sanctuaries to add to existing scattered parks and refuges. "But even after many, many years of that, it was really just addressing a fraction of the need," says Mark Reynolds, lead scientist at the California branch of the nonprofit. The impact on the birds was severe. Surveys showed that migratory birds coming through the valley dropped from 30 or 40 million a season to somewhere between three and five million, according to rough estimates from a decade ago.

That's when the conservancy decided to try something new. As

Reynolds's staff kicked around solutions, somebody asked him a question: "Have you thought about renting habitat instead of buying it? It looks like these birds only need this for a few weeks out of the year." The idea was simple, but putting it into motion took ingenuity—and detailed information. Could they find a way to create what amounts to "pop-up" wetlands only when the birds need them for migration rest stops? Reynolds knew eBird data was getting more sophisticated and ended up working with the Cornell Lab to analyze the intricate timing of the migration routes. California's eBird use is higher than any other state, so the information on when and where migrating birds passed through was precise and plentiful. They called the project "Bird Returns," and put out the word to see if farmers were willing to leave water on their fields—mostly rice ponds at first—for a few extra weeks before draining them.

The projects asked farmers to submit bids for the payments they'd like—a kind of reverse auction that set levels for the incentives. Nowadays one hundred farmers a year are setting aside 55,000 acres along the valley where birds rest and refuel. The initial costs came to about $1.4 million a year—less than 10 percent of the cost of buying the land. The experiment paid off for the birds from the start: Researchers with The Nature Conservancy and the California-based research center called Point Blue Conservation Science found two and three times more shorebirds in the flooded Bird Returns fields. The number and variety of birds at times reached the highest ever reported in the region.

Bird Returns has also captivated many of these farmers. "Now pretty much all the farmers have cards that have all the shorebirds on them," says John Brennan, who manages the Davis Ranches, a 5,400-acre rice farm an hour north of Sacramento that's one of the original partners on Bird Returns. "They can tell you that's a Dunlin over here and that's a Dowitcher over there."

The success of the project has drawn attention from across the country and boosted contributions. This past year, the Bird Returns annual budget hit $10 million. Also, the California Rice Commission and additional private wetland owners came aboard as partners. The expansion has made Bird Returns a key part of California bird conservation strategy, says Greg Golet, an ecologist for The Nature Conservancy and the project's lead scientist. "I wouldn't suggest this is the only way to do it," he says. "The novelty of it has certainly helped people to latch on to it. It can be a heavy lift to get the habitat out there. But it's really worked well."

Another area where citizen science data has made a difference is in determining where the placement of wind turbines is a threat to Bald Eagles. The problem has accelerated as the numbers of both the turbines and the eagles have increased. It's still a federal crime to harm a Bald Eagle, which means the Fish and Wildlife Service must decide if a proposed turbine site poses a threat and whether to issue a permit. To pull that off, the service needed far more detailed data than they were able to find, slowing down the permitting process.

"They approached us to use eBird," says Viviana Ruiz-Gutierrez, the Cornell Lab's assistant director of Avian Population Studies. "They wanted to know, 'Can we trust it?'" So the lab compared eBird data with other bird counts that take readings every year at the same time. Eagles are popular and easy to see, and so the location data from around the country is extensive. More than 180,000 birders contributed eagle sightings over all four seasons used in this research. Jerome Ford, assistant director for migratory birds at the U.S. Fish and Wildlife Service, said the eagle research is probably just the beginning of the agency's use of eBird data for population studies that otherwise require expensive methods such as collecting data by airplane flights. "We're going to rely on eBird as much as we can," he says. "It's credible data, and it's probably much more cost-effective."

The staff at Cornell expects to take eBird in new directions. Director Ian Owens says eBird is now expanding its global reach. It's already in widespread use in Europe, Latin America, and India, but gathering enough data to cover regions of the world, and eventually the entire globe itself, is a tall order. "My hope is that in the next five years, we'll be able to do global analyses for representative species," Owens says. "The next stage is to be able to say how these birds are responding across the world."

Birds as Indicators

On a visit to the National Aviary in Pittsburgh, we came upon a huge photo of a smiling miner, face streaked with coal dust, holding up a birdcage with a canary inside. The bird's only job was to sing and keep that miner alive. Canaries were first put to work in coal mines after an explosion in Wales killed fifty-seven men and eighty horses in 1896. Fifteen years passed before a Scottish medical researcher and mining engineer named John Scott Haldane, who was studying carbon monoxide, found a way to prevent such a tragedy from happening again. Canaries, he discovered, are twenty times more susceptible to carbon monoxide than people, giving him the idea of using the birds to signal when dangerous gases were building up underground. The miners treated the birds like pets, talking and whistling to prompt them to answer back. If a canary stopped singing and fell off its perch, the miners would still have enough time to flee.

Before long canaries were singing in mines around the world and are credited with saving more than a million lives in Great Britain alone. The birds stood sentinel until 1996 when the invention of a handheld carbon monoxide monitor known as "the electric canary" was introduced. Live canaries have been used as detectors in other dangerous situations, too. During the Persian Gulf War, Norman Walker, a British Army doctor stationed in Saudi Arabia, brought his own canary to detect poisonous gas and named him Elvis for "Early Liquid Vapour Indicator System." And after a 1995 terrorist gas attack in Tokyo, Japanese troops in hazmat suits clutched cages of canaries as they stormed a house where the lethal gas was suspected of being manufactured.

We lost count of how many people we interviewed used the phrase "canary in the coal mine," to point out the fact that birds are still warning us about environmental threats. Birds are highly sensitive to the quality of the air, water, soil, and overall climate. When their populations drop, mating schedules slip, or the number of nestlings falls, scientists can often correlate the data and figure out the impacts of climate change, pesticide use, and habitat destruction.

Now birds can help measure the quality of the environments around them. In Sonoma County, California, the wine-growing region just north of San Francisco, nonprofit Point Blue Conservation Science is creating an environmental X-ray of the county's biodiversity. Recorders collect songs from Sonoma's 240 bird species, which researchers then combine with NASA satellite images of the entire county. Putting the ground and space views together reveals a portrait of where biodiversity is high and where it's not.

"What we're trying to see is the biodiversity on a landscape level," says Leo Salas, the Point Blue ecologist leading the NASA-funded project called "Soundscapes to Landscapes." Once the mapping is completed, Salas says they'll look for ways to restore places where the landscape is poor.

Birds are also proving to be indicators of habitat health in the coffee-growing regions of South America, one of the world's most important regions for migratory as well as local birds. Climate change has narrowed the altitudes suitable for growing beans, altered planting and harvest schedules, and may eventually threaten the future of the farms. The CEO of Nespresso, an upscale company that produces pods for espresso coffee machines, visited the Cornell Lab to see how birds could help gauge the health of farms in Costa Rica and Colombia. The research uses a combination of audio records and eBird data to track the more than two hundred species found on many coffee farms, gauging such factors as elevation, tree coverage, insect life, and temperatures.

They've found that where the bird populations are healthy and stable, the environment is best for coffee. Where trees and shade are plentiful, so are birds that can find food and habitat—and that in turn creates more favorable growing conditions. "It's abundantly clear in all of this that if birds are doing well on coffee farms, the population of other animals are likely doing well, too, and that the water is clear and the soil and forests are healthy," said Eric Poncon, regional director for the coffee division in Central America for Ecom Agroindustrial Corporation working on the Nespresso research.

The project relies on eighty audio recorders on the Nespresso coffee farms and eBird data collected by farming staff to track the numbers and types of birds. This tracking helps create the scientific analysis used to sharpen farming practices. "This increases how much coffee they can generate and the quality of the coffee," says Ruiz-Gutierrez, who oversees the project for the Cornell Lab. The results are on display around the farms every day. "These farms are alive with bird song."

She tells of meeting with coffee companies in Nicaragua to talk about how bird-friendly coffee can help them create healthy, sustainable agriculture. At one point, Ruiz-Gutierrez stood with a group of growers on a farm out in the countryside, far from any towns, and asked everyone to listen to the birds. "I said, 'If you close your eyes, you can actually hear sustainability right now,'" meaning that bird songs are in effect proclaiming a healthy environment. "Birds are an excellent indicator for measuring sustainability."

Scott Sillett, director of the Smithsonian Migratory Bird Center, which has pioneered research in coffee-producing countries for decades, says the realization that birds can protect farms at the same time as farms protect birds is spreading. "This is such an important emerging issue, to be able to understand how birds help us," he says. "The reason why we focus on coffee is that the regions where coffee is grown are often the most biodiverse areas. Anything we can do to help make these landscapes less detrimental to biodiversity, to wildlife, is a win."

One of the more futuristic roles eBird may play is to use birds to track the number of pollinators such as bees and butterflies. The Cornell Lab is working on an experiment with Walmart to study whether the makeup of bird populations in a region can indicate the number of pollinators, for instance, on the farms that supply the giant retailer. The Walmart Foundation awarded the lab a two-year, $500,000 grant to test out this idea. Small insects like bees are hard to see, and thus difficult to track. But birds are easily spotted, and the eBird system is already mapping birds almost everywhere.

"We know many birds and pollinators respond similarly to farming practices and habitat conditions," says Cornell's Rodewald. "If they respond similarly to changes in those resources, that would make birds a suitable indicator for the bumblebee." From there, it's a matter of matching what birds are around which insects and butterflies. "Once we've identified the bird species that closely track the pollinator communities, we unlock the potential for birds to be used as surrogates."

The concept is already drawing interest from other companies even while their experiment is still under way, Owens says. "I think groups like Walmart are really feeling that with the next generation of consumers they need to lift up their green credentials," he says. The companies are looking for ways they can measure the health of the environments they rely on in ways that haven't been possible before. "What they want is for us to give them an index that can measure this quickly and easily. They want almost a kind of product that can do this."

Lynn Scarlett, former deputy secretary and chief operating officer at the U.S. Interior Department who's held a series of policy and environmental posts with agencies and nonprofits, sees these projects as evidence that some of the biggest companies are starting to recognize the value of a robust environment to both production and their status. "I think we're beginning to see more and more investment in biodiversity," she says. "It may partly be the optics, or partly because they're beginning to see that, oh my goodness, nature matters, birds matter."

Either way, Michael Doane, who travels the world for The Nature

Conservancy working with agribusiness firms, says that corporations are getting a strong message from consumers and investors. "There's not a CEO in agribusiness that doesn't have to deal with talking to investors or employees or customers about these issues," says Doane, who grew up on a Kansas farm, worked at the agrochemical company Monsanto for a decade, and now is the conservancy's global managing director for food and freshwater. "They have to have a good plan for where they're going."

Birding as a Game

When another year of eBird information is released in bulk each January, Scott Stafford just can't help himself. A compulsive multitasker with great data skills, he creates a twenty-two-page report for Washington-area birders that shows who's seen the most birds for the year in categories—by the month, by rare birds, common birds, for rookies, in all four seasons—just about every way you can look at who's tops in the search for birds.

Stafford's annual summary illustrates an aspect of modern birding that helps make eBird so popular: the competition and gamesmanship made possible by connecting hundreds of thousands of birders and their lists. The eBird staff realized early on that competition helps motivate its contributors. Every month the lab posts the names of top eBirders by local county, state, country, and world, both by the number of species and total checklists they submit. There's even a monthly drawing for a pair of expensive binoculars.

We were curious to know how much effort it takes to get to the very top of the world's eBird list. A lot of time flying, says a retired diplomat who spends about half the year traveling in search of birds he's never seen. Peter Kaestner has logged as many species as anybody else on earth. That now stands at 9,727 at last count, just short of the world total. The Baltimore, Maryland, resident has had a couple of advantages. First, he studied biology and ecology, giving him a wealth of knowledge about birds. Then he joined the U.S. Foreign Service, which over the years posted him in India, Colombia, Brazil, New Guinea, Malaysia, Namibia, Guatemala, and the Solomon Islands—near many of the hottest birding spots there are.

Everywhere he went, he took side trips into the rainforests, deserts, swamps, and mountains to look for birds. Early on, in 1986, he became the first person ever to see species from each of the 249 families of birds. While posted in Bogotá, Colombia, Kaestner uncovered a rare antpitta, an olive-brown bird with dull, mottled plumage unknown to science until he recorded its song and got its photo on the eastern slope of the Andes. As eBird's appeal started to grow, he managed to stay ahead of the pack. "I love the competition," says Kaestner, now approaching 70. "I love seeing my name on the top of the list. I love the idea, you know, that I've seen all the families, and if somebody created a new family that I haven't seen, I'd be on the next plane to make sure my list of families is complete."

Kaestner's motivation isn't limited to just ticking off birds—a criticism leveled against many world-traveling listers drawn to the sport of chasing rare birds. He's deeply concerned about the loss of species that parallels his years as a birder. "What's happened during my lifetime is very sad," he says. "We are doing things to the environment now, the scope of which were done in the 1800s, but we're doing it much more insidiously because we're doing it with chemicals and pollution, with climate change that is much more difficult to rectify."

One of his most valuable contributions to eBird is reporting birds in places around the world few others venture. "I really, really like the idea that I'm contributing to something greater than myself," he says. "Every time I fill out an eBird checklist, I know these data are going into a database that is used for scientific purposes and to better understand the world. That's core for me, that these contributions are a legacy. They'll be there forever."

Postscript: How Photography Is Helping Birds

From Anders:

I was mesmerized the moment the Great Blue Heron stepped into my viewfinder on the edge of a pond on Mississippi's Gulf Coast early in our year of travel. The heron was in the full bloom of his spring breeding plumage, with long strands of feathers curling down his neck and back. His beak had changed to a bright orangish yellow. Just over four feet in height with a six-foot wingspan, Great Blues are hard to miss any time of year. They're found in every part of the continent, often posing like statues in shallow water waiting for the right instant to snag a fish. That helps make them one of the species most often reported by eBirders, which is why you see a Great Blue Heron's photo at the top of this chapter on the evolution of birding tools.

Several chapters earlier, the featured species is the tiny Florida Grasshopper Sparrow, which we've been following for years as scientists try to restore their depleted numbers. For most of our visits on that project, the sparrow was a phantom, hidden in its aviary or skittering out of sight along the base of the grasslands. But then came the March morning on the Florida prairie when a fresh generation of sparrows went looking for mates just after sunrise. Suddenly, the birds transformed into miniature performers like the one you see at the top of Chapter 1. This wasn't an easy photo, since the fog was still heavy and the birds kept moving from one spot to the next. But one brave sparrow stuck it out long enough for the mist to ease and let me come within twenty yards, just long enough to get the shot.

There's a story behind each of the photos at the top of these chapters. Our travels were mostly devoted to searching out the people behind the rescue missions and experiments to save birds. But the visits were also about getting the chance to watch these symbolic species in action and catch their photos in ways that say something about their existence. The pursuit came with all sorts of lessons about studying and understanding birds.

Starting about a decade ago, I dusted off a few basic photo skills I'd picked up during my first jobs as a newspaper reporter. I knew photography had changed dramatically over the years with the conversion to digital technology, but I was still swept away by how the ultra-sensitive cameras and precision lenses can make a bird photographer out of almost anyone. I love to spend my days moving in silence along grassland trails, marshes, and woods to watch a bird nesting, foraging, or suddenly taking flight. I've found myself part of a wave of hobbyist photographers drawn to the challenge of birdwatching through the viewfinder. Many days, cameras are as common as binoculars on nature trails. Just as the millions of eBird posts help provide the foundation for research, the steady flow of bird photos helps capture the beauty and wonder of birds in the wild. Our collective output fills websites, social media, and citizen science projects like eBird with photos that help put birds on display. More than 200,000 photos of Great Blues, for instance, have flowed into eBird's galleries from around the hemisphere.

The best wildlife photos can sometimes change the course of events. A century ago, in the earliest days of wildlife photography, the stunning work of pioneers William L. Finley and Herman T. Bohlman helped convince President Teddy Roosevelt to create a network of wildlife refuges to protect birds and expand the national park system. These days, *National Geographic* photographer Joel Sartore uses his photography in a similar pursuit. He hopes to motivate people to care about troubled species with his project called the Photo Ark that's led to a series of magnificent books as well as scores of galleries loaded with birds. He's spent a quarter century making his way around the globe taking photos of every one of the 20,000 species living in the world's zoos, parks, and refuges.

Bird photography turns out to be about much more than the photos. It's a pastime that teaches a graduate course in patience along with a minor in concentration. You have to understand the species and its habits to capture its moves that unfold in split-second intervals. Best of all, you get to study these creatures through lenses that magnify the birds by ten to twelve times, until you can see the threads of their feathers, freeze a bird in midflight, and peer into the nests of the smallest hummingbirds. It's the best way I've found to grasp what the birds are telling us. And on those lucky days if you're there at the right time, a bird such as that heron in full breeding glory will step into your path for a moment or two.

Teams of researchers in both South and North America are trying to protect the Cerulean Warbler, one of scores of tiny world-travelers whose ranges span the hemisphere. This sky-blue bird, like many of its migratory peers, is one of the toughest species to restore.

7

WORLD TRAVELERS

Mindo, Ecuador

For Cerulean Warblers, the year follows a predictable pattern. By late May or early June, most of these compact, baby-blue birds have finished yet another migration and arrive at their summer homes in the Appalachian Mountains. Here they'll settle on a mate and build a nest that would make an engineer envious.

First the birds choose a suitable tree, a towering white oak or perhaps a cucumber magnolia, but it must be in a forest with gaps separating one leafy canopy from the next. The industrious warblers then spend the next five days gathering precise construction materials: grapevine bark fiber, fine grass stems, some horsehair if they can find them, or even damp pieces of newspaper. All this gets woven together with spider or caterpillar silk into a teacup-like shape that's exactly two inches tall by two and a quarter inches wide. It doesn't take much to hold three or four oval eggs the size of small marbles.

For the next couple of weeks, the female lays her eggs and sits on the nest until the chicks hatch. Just ten days later, the little Ceruleans perch up on the edge of the nest and learn to fly. Toward the end of July the entire family is ready to embark on a miraculous 2,500-mile journey to spend the winter in South America.

They first fly to the Gulf of Mexico, eating enough insects to double their weight and storing enough energy for the six-hundred-mile, non-stop crossing to Mexico's Yucatán Peninsula. Once the Ceruleans hit land some eighteen hours later, they proceed to one of several countries along the top of Latin America, arriving sometime in late September.

In northern Ecuador, George Cruz is waiting for the Ceruleans to show up in his slice of the rainforest. It's a sign that a flood of birds from North America are headed his way. Cruz breaks into a big smile as he talks about how the Ceruleans blend right in with the local birds. "Such beautiful blue birds," says Cruz, a veterinarian who gave up his practice to build a tourism business centered around birds. "People love to see them."

Ceruleans are just one of more than fifty types of small, colorful warblers, and like many of their cousins, the Cerulean's conservation on an international scale is awkward. These birds are like the offspring of divorced parents who don't see things the same way and blame each other when things go awry. North and South America share an estimated 350 long-distance migratory species, and many of them are in decline. Despite years of good intentions, governments and nonprofits in North and South America have difficulty coordinating their conservation efforts in a way that sustains these "neotropical migrants," as they're called. "This is a global problem," says Paul Greenfield, a longtime resident of Ecuador and bird artist who's helped create several bird foundations. "All of these organizations should be working together."

There's a lot to do. With 1,659 species, Ecuador has the highest number of birds per square mile of any South American country, making it a popular tourist destination. Most of Ecuador's birds are part of an exotic, year-round lineup of toucans, hummingbirds, cuckoos, and condors. Even so, the part-time Cerulean Warbler is valued as rare, says Cruz, who operates bird preserves and owns six tourist lodges here.

Ceruleans have been singled out for rescue in both North and South America because they've lost population at the steepest rate in the warbler family—about 70 percent over the past fifty years. The Cerulean's case is especially interesting because even with such a sharp rate of decline, there are still more than half a million birds left—enough to make recovery a real possibility. The efforts to preserve and restore Cerulean habitat in both Ecuador and the Appalachian Mountains tell a broader story of local conservation work needed to keep songbirds and other migrating species stable. The alternative is these songbirds will dwindle away until there are so few, the only way to save them is an Endangered Species Act listing.

Roughly two hundred U.S. migrating species have been in this gradual decline state since at least 2008, according to the U.S. Fish and Wildlife Service, yet their numbers are still high enough for interventions. If the current Cerulean habitat protections succeed on both continents, the bird could illustrate the benefits of early action to preclude an avalanche of extinction risk. The work in both South and North America is complicated and costly, but nothing like the herculean effort required if the Cerulean were to make the endangered list.

"It's Like My Head Is Exploding"

We came to Ecuador because the country reflects the challenges of the entire continent. All the topographies of South America are tucked into one country the size of Colorado: Amazon rainforests in the east, the astounding biodiversity in the country's northern half, and the Andes mountains that run down the center of the country like a backbone. The Galápagos Islands, where Charles Darwin did his evolution research with the help of the families of finches, are six hundred miles off the coast. All of this makes Ecuador one of the most biodiverse nations in the world. But like all of South America's twelve countries, Ecuador struggles to balance economic and environmental needs. Even though an impressive 20 percent of its land is protected, Ecuador leads the continent in deforestation. The government has launched some progressive policies, such as small incentives that pay landowners to protect their forests. But the country's leaders traditionally give preference to mining, logging, and agricultural interests over the environment.

Ecuador, like Latin America as a whole, is absolutely loaded with birds. On our third day here, Juan Carlos Crespo is the first to jump from the van as the sun comes up along a dirt road in the Mindo region in the heart of northern Ecuador's richest territory for birds. He literally turns in circles as he takes in the morning symphony of bird songs that come from every direction. "Oliver-crowned Yellowthroat up in the weeds," he says. "Whiskered Wren on the hill. Look at the top branches toward the back." He's pointing here, then there, sounding off like an auctioneer. "Blackburnian Warbler," he adds, and then goes into a list of even more birds that suddenly surround us, including parrots, woodpeckers, tanagers, and warblers flying this way and that along the road.

Crespo is our guide on a trip around the countryside arranged by the Jocotoco Conservation Foundation, Ecuador's leading bird nonprofit. We're tagging along with Mike Parr, president of the American Bird Conservancy, and Dan Lebbin, the conservancy's vice president of threatened species, as they work with Jocotoco to size up possible land purchases. ABC has been one of the most active U.S. entities helping fund and protect lands in Latin America. The conservancy has a clear aim: working with partners to make sure that at least some habitats are in place for the five hundred most threatened birds in the Southern Hemisphere.

For now, though, Parr and Lebbin are distracted. Both are obsessed, world-traveling birders. Lebbin has made more than fifty trips to the continent. We stop along the roadside, two hours northwest of the

The Crimson-rumped Toucanet is part of Ecuador's hugely varied birdscape that gives this small country more species per square mile than any in South America.

capital of Quito, to see what birds we can find. So many appear—mostly year-round species with a few migrants mixed in—that it's hard to keep up with them. "It's like my head is exploding when it's like that," Crespo says later of the morning's bounty.

The trip shows off the unparalleled biodiversity in this part of Latin America. Though it's a small country, Ecuador has more bird species than all of North America combined. It also has more than 350 species of mammals, 1,250 types of fish, and more than 15,000 species of plants—2,725 of them orchids. It's home to the world's biggest collection of hummingbirds, 130 varieties, some with long trailing tails, others with neonlike colors, and a few whose beaks are longer than their entire bodies. The forests are full of toucans, those comical, rainbow-colored birds with oversized bills. In the mountains, you find Andean Condors, whose ten-foot wingspan makes them the largest birds of prey in the world. With such a dizzying year-round wealth, the visiting Ceruleans we're following are easily overshadowed. That's partly because they're such elusive birds, hiding behind leaves high in the canopy.

Protecting the Cerulean Warbler is especially complicated because it needs quality habitat in all of its range—science-speak for all of the places it either lives or migrates through. And the Cerulean's range is massive on both continents, and so is habitat loss, its main threat. The bird's winter range covers more than a third of northwest South America from Panama, Costa Rica, and Venezuela, down through Colombia to Ecuador and parts of Peru. The area where they breed in North America is also huge. Roughly 80 percent of Ceruleans settle in the Appalachian Mountains for the spring and summer, but they also spread from Minneapolis to parts of Canada and into every part of the upper East Coast. "We can safeguard the endangered endemics [native local birds]. But the migrants spread across multiple countries," says Parr, referring to the massive territory birds like the visiting Ceruleans cover during their winter stays. "This is one of the biggest challenges for us."

Both halves of the hemisphere have taken different conservation approaches. In North America, the Cerulean has been the subject of plans, debate, and false starts for decades. Only recently has a consortium of state, federal, and nonprofit organizations come together with a strategy that deputizes private landowners to try and help the Cerulean.

Reconfiguring Forests in Virginia

Jefferson National Forest, Virginia

As we follow biologist Todd Fearer through the woods in southwestern Virginia, he gets excited when we reach a stand of giant oak trees in the Jefferson National Forest. The tree canopy looms fifty feet above us, perfect for Cerulean Warblers. But Fearer starts pointing to the ground scattered with ferns and brush plants. This is where a lot of the benefits of his rescue work become apparent. He's especially proud of how the pine sprouts, which the Ceruleans don't like, are thinned out. But we keep getting distracted by all the Blue Jays, robins, flycatchers, and sparrows. As the September afternoon light filters through the branches, we see why the birds are zooming around. Clouds of insects look like an airborne cafeteria.

At this point in the year the Ceruleans will just be arriving in Latin America. But when they return next spring, Fearer says these woods will be groomed to meet the warblers' every need for breeding. The Jefferson National Forest is part of an $8 million conservation experiment covering 12,000 acres of mid-Atlantic forestland. The project is being led by the Appalachian Mountains Joint Venture where Fearer is director. Joint ventures are public-private partnerships that work on broad-scale problems where birds face challenges over multiple states.

The problem for Ceruleans is that many Appalachian forests were almost entirely harvested more than a century ago. As the trees grew back, the forest turned into a uniform habitat Ceruleans struggle to adjust to: The trees were mostly the same height, and the stands were crowded with invasive species. The forest lacked the naturally occurring gaps in the canopy that make it a healthy ecosystem for the Cerulean and other birds.

He points out places where workers have removed the invasive plants and trees, thinned out crowded stands, and replanted native trees. Pesticides are banned, and thus the smorgasbord of tasty insects. If the work proves successful at bringing in more warblers over time, this track can serve as a model for broader efforts elsewhere for Ceruleans—and other forest wildlife.

Fearer's first challenge was to find enough forestland available for restoration and enough money to pay for the work. Although national forests and other public lands weave through the Appalachians, most tracts are owned by individuals. "A good 75 to 80 percent of this land is in private hands," he says. "So if we're going to make a difference with the Cerulean, we've got to focus on those private landowners."

Ceruleans Are a Classic "Carrot-Stick" Scenario

Both private owners and the government have some of the same incentives to cooperate on a bird like the Cerulean—it's in a middle ground of declining but considered still healthy enough to recover without federal protection. It's more cost-effective for the government to fix a bird population before it gets to the brink of extinction, and it spares the landowner land use restrictions that can come with federal protections. For U.S. landowners working with Fearer's Cerulean Warbler project, the Department of Agriculture draws on the massive agricultural funding legislation called the Farm Bill to reimburse 75 percent of forest improvement costs that help the cause.

The Cerulean is on the radar screen because it already brushed up against the Endangered Species Act in the year 2000. By that point the bird's losses were so precipitous that twenty-seven environmental groups petitioned the Fish and Wildlife Service to list it. The request came as no surprise; since 1966 Cerulean numbers had been declining by about 3 percent each year. But state and federal wildlife agencies hadn't been able to decide what, if anything, to do about it.

"Honestly, there were a lot of state wildlife directors at the time who never thought about these birds," says Bob Ford, a veteran state and U.S. Fish and Wildlife Service biologist who's also a migratory bird expert. "They thought these birds would take care of themselves. They didn't need special habitat or anything." Many local wildlife agents were focused on game species and simply didn't consider songbirds a part of their job. Some of them didn't even realize songbirds used the surrounding property. "They thought songbird management meant putting up bluebird nest boxes," Ford says.

Once the Cerulean's petition was filed, the Fish and Wildlife Service would need to move quickly to meet its legal obligations. The law gives the service fifteen months to make a final listing decision, and an Endangered Species Act ruling can be tricky because the criterion is at once precise and vague. A species makes the ESA list when it's in danger of extinction "in the foreseeable future" everywhere it lives (its range) or in a "significant portion." Specific parameters of these terms, like "foreseeable future," are not defined by the ESA.

In the case of the Cerulean, the request was for a slightly less restrictive listing called "threatened." This criterion is even more vague: A "threatened" bird is *likely* to become" an endangered species in the foreseeable future everywhere it lives or in a *significant portion* of its range.

For a bird that flies as far as the Cerulean during migration—some 2,000 to 3,000 miles each way—that leaves room for interpretation. If a group of the warblers is healthy in at least some places, that's not all of its range. When Fish and Wildlife denied the Cerulean protection in November 2006, that was the rationale. The service acknowledged that in some places it lives, the bird is likely to be at risk of extinction in less than a hundred years. But they didn't consider those places significant enough to put the species at risk of complete extinction.

The Cerulean's denial came a full four years past its fifteen-month deadline. This was no surprise, either, because the service had long been underfunded compared to the increasing demands. By 2011 the problem reached federal court when the number of past-due ESA rulings had grown to 750. The lawsuit settlement gave Fish and Wildlife five years to take care of the backlog, but by the deadline, 417 species were still waiting.

Today the problem is so severe the service doesn't even pretend it can meet its legal obligations. The service still gets ninety days to decide if a petition warrants the full twelve-month review needed for a "yes" or "no" ruling. It has resorted to publishing a five-year work plan designed to help the public "predict" when a species will get high enough on the waitlist for review. But even those predicted dates are often missed. The first order of business in 2022 was to resolve the forty species with court-ordered deadlines. Meanwhile, hopes for the backlogged birds and other wildlife and plants continue to fade.

Fortunately there are other avenues for bird advocates to pursue. Experts and others sometimes form working groups to study a bird, but members are volunteers and must find funding on their own. In 2001, such a group of scientists, government staff, and folks from coal and forest industries zeroed in on the Cerulean to make specific conservation recommendations. Over five years the group obtained grants to continue studying the bird, met with coal industry leaders to discuss habitat issues, and traveled to Ecuador and Colombia to talk with local groups about the threats in South America.

In 2007 the group laid out a roadmap of actions to stop the declines and then double the Cerulean's population over the following fifty years, an ambitious and probably impossible goal to meet without significant additional investment. The group mostly pinpointed additional research gaps, and simply stated that habitat improvements were needed everywhere the bird lived. Fast-forward to 2022, where the joint venture is a hopeful example of the sort of work that needs to continue throughout the Cerulean's range.

The day we tour the forest with Todd Fearer, the first experimental phase of his project is coming to an end. It's hard to document progress from the work, since their 12,000 acres is a mere fraction of the tens of millions of acres in the bird's range. But Fearer thinks the results are sufficient to attract more funding, and he hopes to restore enough forest to forge a safe pathway for the birds from these breeding areas to the Gulf of Mexico. That's the jumping-off point for the Cerulean's semiannual migration to the flip side of its annual range. In Ecuador, the warbler joins thousands of other birds vying for a place in this crowded biodiversity capital where they spend their winters.

Protecting the Chocó Corridor

Canandé, Ecuador

Martin Schaefer holds the wheel of his Toyota pickup as if gripping the reins of a rodeo bull as he nudges his truck along a logging road near the Colombian border. Schaefer is head of the Jocotoco Conservation Foundation, and he's taking us to see property he's evaluating for possible purchase. We're at the heart of the Chocó Corridor, the multicountry swath running along South America's western edge that's one of the most important stretches of forest for wildlife. Nineteen types of monkeys, jaguars, and sloths as well as eight hundred species of birds including the Cerulean live in these forests.

We pass logging camps, poverty-struck shanties, and cacao farms, and the drive becomes a painfully slow passage over boulders and through muddy ditches. "I feel like I'm in a washing machine," says Dan Lebbin from the backseat. He's working with Jocotoco on the purchase in his role overseeing threatened species for the conservancy. The rush is on to get land for preservation in the Chocó Corridor, especially in Ecuador's portion where only 2 percent of the original vegetation remains.

Every half hour or so Schaefer pulls to the side to let one of the heavily loaded logging trucks slip by within inches on the narrow roadway leading to market. "Most days, we see twenty to forty of these trucks pass. Some days, it's as many as sixty," Schaefer says. "You can literally see the forest disappearing every day."

Once logging companies take the largest hardwoods, local squatters move in and remove many of the remaining native trees and bushes to plant cash crops—cacao, vegetables, fruit, palms, and balsa wood. The international markets for these products are constantly shifting, not always for the better. When the wind industry needed lightweight wood to make turbine blades, for example, the demand for balsa boomed. But now the wind industry has shifted to other materials, and a fungus decimated palm trees grown for oil. Where bamboo and palm trees replaced rainforest, remnants of these ruined crops stand abandoned along the road. "So everybody loses out," Schaefer says, "and the environment degrades in the process."

When Schaefer tries to get started after letting the latest truck pass, we begin to slide sideways down the sharp hill toward a gully and stream. He nudges the pickup forward and back, and slides even farther down before finally gaining traction. This makes his passengers more than a bit nervous, but Schaefer seems to think nothing of the prospect of rolling into a gulch miles from where we began that morning. Treacherous roads are just one of the reasons Ecuador's rainforest can be a hazardous place to work. Not only is it populated with jaguars, poisonous snakes, and scorpions, negotiating to buy land in a place with opponents of conservation has its dangers as well. Members of Schaefer's staff are sometimes threatened at gunpoint. "This is a lawless area to some extent," he says.

That's why Jocotoco is trying to do more than just set aside land for birds. The foundation is working to create jobs in local communities, many of which are home to Indigenous tribes. "You can't do successful conservation if you're not thinking about the people that live around you," Schaefer says. "At some point you have to step up to these responsibilities and find models that allow people to make a living."

These days Schaefer's mud-caked pickup is a familiar sight, and people wave and honk in greeting. Most everyone knows the foundation is in the market to expand and protect the corridor, and Jocotoco's staff waits for sellers to come to them. Although some people don't seem to care who buys their land, Schaefer says the number who support rainforest protection along the Chocó Corridor is growing. One landowner who had agreed to sell to a neighbor found out his buyer had contacted a logging company. Loggers had already marked trees to cut before the sale went through. "So the owner said, 'If you just want to buy my property to cut down the trees, I'm not selling to you,'" Schaefer says. "'I'm selling to the conservation people. I'm happy to see that the land I owned will remain.'"

Tourists Do More than Just Watch Birds

One of the ways local landowners are able to keep their portions of the rainforest intact is by setting up bird parks for tourists. Only about 15 percent of the original rainforest in Ecuador is still standing, so every parcel is precious. Angel Paz was eking out a living on a small farm in northern Ecuador, when he saw an unusual bird with a large, pointed beak and dark bulbous eyes. With a worm in its mouth, the bird made only the briefest appearance before fleeing into the woods.

The bird turned out to be a Giant Antpitta, one of most elusive species in Ecuador as well as Colombia. Paz's glimpse that day gave him an idea. "I began to follow the bird into the cloud forest," he says, "offering her worms." Weeks of patience and dozens of worms later, the wary antpitta finally began accepting Paz's offerings. It was the start of a relationship that led Paz to a profitable career in tourism.

Over the past decade tourism has been growing as an important piece of the conservation formula for birds in Ecuador. Tourists pay to see exotic birds in the wild, and small landowners like Paz make a good living while preserving their land. "There's a sort of snowballing thing once you get people interested in birds, and I think that's unbelievably important," says Paul Greenfield, the Ecuador bird artist and expert who has also helped train bird guides in Indigenous communities.

Privately owned bird parks like the one Paz and his brother built, called Refugio Paz de las Aves, also help visitors learn about conservation in Ecuador and foster an appreciation for wildlife. The Paz brothers began by building a series of trails to see the Great Antpitta and to cliffs where colorful tanagers roost. Another trail led to mating territories of the Andean Cock-of-the Rock, foot-long, bright-red birds that put on a riotous dance for females. The brothers worked for months preparing for tourists, but nobody showed up. The paths were growing over when a lone birder arrived and spent four hours on their farm. "We were paid $10, which was the equivalent of a complete day's work for a farmer," Paz says. "We were surprised and overjoyed."

Angel Paz's farm outside the small town of Nanegalito is an unlikely tourist destination. The sixty acres he inherited from his mother sit on a ridge with slopes that drop nearly straight down on either side. Many of his neighbors live in wooden shacks, covered with tarps to keep out the rain; their yards are home to underfed chickens, cats, and dogs. Dairy cows, a central part of farming here, roam at will along the hills. It's hard not to get caught behind a small herd heading down the mountain to get milked.

Paz's complex stands in sharp contrast to the neighboring countryside. Paz has now built a brand-new lodge, hummingbird gardens, pathways around the property, and a restaurant where his wife, María, prepares the meals. Rates range from $35 for a half day visit to $290 for three days including tours, lodging, and meals. In a good year, he'll get 2,500 visitors.

The bird park formula works, Greenfield says. "People think, 'I could do that.' You put up a hummingbird feeder and a couple of bananas to attract the birds," he said. "You don't have to be a genius to do it."

In the two decades since Paz got started, bird tourism in Ecuador has expanded to include a small army of local guides, an array of locally owned lodges, and stand-alone bird gardens. International tour companies bring in well-to-do birdwatchers from around the world. "[Bird tourism] is just all over the place in Ecuador right now," Greenfield says.

Climate Change Funding Helps Birds, Too

South America has struggled for years to get conservation support from wealthy foundations in the North. But that's starting to change as some of the world's biggest philanthropies, including Jeff Bezos's Earth Fund, Mike Bloomberg's foundation, the Gordon and Betty Moore Foundation, and the Rob and Melani Walton Foundation, are promising billions to protect land around the world to combat climate change. These initiatives are still getting started, but among the Earth Fund's first grants are projects to help preserve land in the Andes, the Galápagos, and the Amazon.

The Chocó Corridor is one of the areas singled out for more protection, and one of the keystone birds that would benefit is the Cerulean Warbler. The flow of conservation funds exceeds anything before it, says Byron Swift, an environmental lawyer and senior advisor to Re:wild, a nonprofit founded by a group of conservation scientists and actor Leonardo DiCaprio. Swift is reminded of the increase in support for rainforest renewal a few years ago and notes the current boom is driven by concern about both biodiversity and climate change. "But this increase in funding is already much greater than what we saw before with rainforests," Swift says. "I think people are aware that [climate change] directly affects their survival."

Another venture to preserve land starting at the same time is aimed directly at protecting birds in South America. The National Audubon Society is leading a project to purchase 300 million acres of critical land for birds in partnership with BirdLife International, American Bird Conservancy, and a network of South American environmental funds called RedLAC.

Leading the project is Audubon's Aurelio Ramos, who grew up in Colombia, studied economics, and has worked on environmental issues in Latin America for thirty years. His first assignment at Audubon was to find a new bird conservation strategy for all of South America. "We started with a blank sheet of paper," he says. "What should we be doing, and who should we be doing it with?" These questions led to an analysis that singled out the most critical bird territories on the continent, which turned out to be hundreds of millions of acres. The consortium named the project Conserva Aves, and aims to preserve the first 25 million acres over the next five years. "It's a huge area," Ramos says. "It's just overwhelming, but that's what the science is telling us is needed."

Conserva Aves takes aim at local as well as migratory birds to balance out the needs of species from the North and South. Another part of the initiative is striking a careful balance among the diverse set of partners and coordinating groups that haven't always gotten along. "One of the first things I heard when I came to Audubon was, look, BirdLife and Audubon and ABC, they have their history of competing with each other," Ramos says. "So when I first came, some of the donors said, 'Aurelio, your first task is to get these organizations working together.'"

Leaders of Ecuador's bird groups are encouraged by how the contributions and initiatives clearly recognize the importance of South America for global conservation. "I see North American organizations much more open to understanding the linkage between the North, the Central, and the South of the continent," says Itala Yépez, head of conservation for BirdLife International's Americas Team. Greenfield, the veteran of Ecuador's bird circles, puts it this way: "I remember early on in the U.S., people would say the problem with migratory birds is in

South America, that we were the ones killing these birds. But at this point, in a sense, we may be the key to saving them."

Is Congress Paying Its Fair Share for Bird Support?

Our migrating Ceruleans, like kids bouncing between two homes, need financial support from both parents. But what happens when one side just can't pay? Migrating songbirds are getting more attention in the United States since the Three Billion Bird report showed how hard they've been hit. But when it comes to the big problems on the wintering grounds of Latin America, most lawmakers would rather not get involved. "It's always been more difficult to get funding that goes outside the U.S.," says Lynn Scarlett, the former deputy secretary at Interior.

A primary source of U.S. funding specifically to help protect the more than 350 types of birds that migrate to Latin America comes under the Neotropical Migratory Bird Conservation Act. For years the annual funding held steady at around $4 million, a pittance compared to the hundreds of millions just one high-profile species like the Greater Sage-Grouse gets in the American West. To make matters more difficult, the money is doled out in small grants that require Latin American governments, nonprofits, and land trusts to match the funding three-to-one. Many organizations just don't have the cash to qualify.

But there is hope. The 2021 allocation for the Neotropical Act was increased to $6.5 million, and a proposal to raise it to $20 million was moving through Congress in late 2022. Even so, the American Bird Conservancy's Mike Parr says the urgent need is far greater. "It should be $600 million with no match," he says. That level of funding is not likely, say those who've struggled to increase the support for the birds both hemispheres share. "If you have dollars going out of the U.S. to foreign countries or organizations, Congress's attitude becomes much more penny-pinching and frugal," says Swift of Leonardo DiCaprio's Re:wild nonprofit. "It doesn't play well politically."

Postscript: Completing Our Own Cerulean Cycle

From Beverly and Anders:

High Island, Texas

Our first meeting with a Cerulean Warbler was both exhilarating and frustrating. Late in April a few years ago we got an email alert from a local birding group: A male Cerulean was hanging out at a bridge near our home in Raleigh, North Carolina. He'd somehow veered off his usual migration route, and it caused a stir in local birding circles.

The Cerulean, like all fifty of the tiny wood warblers, is exquisite and often hard to find. By the time the Cerulean arrived in Raleigh, we'd been searching in vain for two years in five states for a glimpse of this bird. He may as well have been the Loch Ness Monster. So when that text alert arrived, we dropped everything and headed for the bridge.

The odds were in our favor. It was the beginning of the breeding season and the males were singing. Ceruleans tend to hide at the tops of very tall trees, so listening is the best way to locate one. As described in our guidebook, the Cerulean sings "a buzzy short series of low-paired notes, followed by a mid-range trill and upslurred, high-pitched *zhree*." This is a lot for untrained ears to digest and, as we hiked toward the bridge, we played a recording to let the song sink in.

When we arrived, a group of birders was already scanning the tree-tops. After an hour Beverly got antsy and wandered down the trail. Suddenly, there it was: a heart-stopping trill, an upslur, and the *zhree*. She phoned Anders. "I think I hear it," she whispered. He alerted the others, and the pack approached in what looked like a sprint in slow motion so as not to scare away our target. "Yep, that's him," one of the other birders confirmed. "Good spotting."

The Cerulean is nicknamed "Sky-blue and Sky-high" for a reason, and all we got that day was a view of his white underbelly, a four-second glimpse of blue, and three blurry photos. We'd heard the best odds to see this sky-blue bird up close is during the spring at High Island near Galveston, Texas, so we decided to include a visit in our research route.

High Island is the welcome center for North America's most flamboyant species—warblers by the bucketload and scads of other feathered beauties, too. It's located on the Bolivar Peninsula on the state's southeast coast. At thirty-eight feet above sea level, High Island is a geographic anomaly—a ridge rising from the surrounding salt flats, the highest coastal landform between the mouth of the Mississippi River and southern Mexico. It's also the nexus where these migrators first hit land after the nonstop flight across the Gulf of Mexico from the Yucatán Peninsula.

Eventually the songbirds will disperse to breeding grounds as far as Alaska, but first, hundreds of thousands of them pause at High Island to recover. A forest-loving bird who's used its last drop of energy will naturally land in the first available tree. Since the trees on High Island are pretty much the only sizable ones along the coast, this is the magic spot. At the height of the migration return trips from South America, the berry trees drip with Rose-breasted Grosbeaks and Scarlet Tanagers, stuffing themselves as if they'll never eat again. The gluttonous spree stains their beaks a berry red; they've hardly swallowed, and it's on to the next bite.

The Houston Audubon Society manages a four-sanctuary complex here, and a $30 donation buys access for the entire migration season between mid-March and mid-May. If you aren't yet a birder, a few days here will surely make you one. You can get a front-row view from bleachers located near an oversized bird bath where birds caked with sea salt shower off. Birders from all over the world converge here, too, quietly pointing out first one species, then the next. It's like being in an art museum where two-legged paintings hop around.

Extensive trails wind through the complex, and benches are strategically scattered around the properties. In 2021 at its Smith Oaks Sanctuary, Houston Audubon opened a new elevated walkway through the canopy of an oak grove. At eye level with the treetops, the walkway is designed so anyone and everyone can enjoy the birds, no binoculars or expertise needed. The idea for the walkway came about after the Houston Audubon staff gathered on the roof of one of their buildings in the late afternoon so they could watch the migrators arriving in the canopy right at eye level. They decided to offer the view to everyone.

We spent several days wandering the preserves around High Island. We saw dozens of spectacular birds, but as luck would have it, the Cerulean did not appear. On our last afternoon, rain mixed with brisk winds came pelting down, and we nearly took a nap. But we knew nasty weather here could force even more birds than usual to land, so we took one more drive along the coast. Before long, we came to a deserted campground in Galveston Island State Park, put on rain jackets, and walked toward the beach.

Sure enough, we saw a mixed flock of warblers landing in some low bushes along the roadway. And there he was, our feisty world traveler, out in the open, shaking off his salty wings. That's the bird you see at the top of this chapter. Our Cerulean looked a little beaten down, but he'd survived the sojourn from South America one more time. In just a few weeks he'd be back in the Appalachians, looking for a white oak tree, collecting grass and horsehair, starting the nesting cycle all over again.

A symbol of Hawai'i, the 'I'iwi is one of the remaining Hawaiian honeycreepers that scientists are struggling to save from disease, loss of habitat, and predators.

8

WHEN ALL ELSE FAILS

Kaua'i, Hawai'i

The plot soon unfolding high in the mountains of Hawai'i is filled with suspense, drama, and sex. The central characters are on a mission to save some of the last native birds from a mosquito-borne malaria that's decimating a dozen of the most precious birds on the islands. But can they reach their target before time runs out? This isn't a movie script. It's the real-life story of the most ambitious bird rescue in the United States that's nearly ready for launch after nine years of preparations. Our hero is a six-legged *Culex* mosquito, and that's where the drama and the sex come in. Scientists will breed tens of millions of the male insects in a warehouse on the mainland, inject them with a bacteria called *Wolbachia*, then ship them off to Hawai'i in low-temperature containers to slow their metabolism for the journey. Sometime in 2024, drones will release them in the high elevations of Maui, and then Kaua'i, where the last of Hawai'i's endangered forest birds live. If all goes according to plan, the sheer number of *Wolbachia*-infected males will overwhelm their wild counterparts and mate with unsuspecting female mosquitoes. The bacteria should act as a kind of birth control, and the eggs will never hatch.

Hawai'i is the extinction capital of the world. Nearly 100 of Hawai'i's 140 native bird species have already disappeared. In the thirty-six years since the Florida Dusky Seaside Sparrow became the last bird to die out on the U.S. mainland, Hawai'i has lost fifteen birds, including eight soon expected to be declared extinct by the U.S. Fish and Wildlife Service. Extraordinary tactics like the mosquito project simply reflect the magnitude of problems raining down on the most remote islands on earth. We needed to see it for ourselves, so six months into our travels, we parked the Airstream in Northern California and flew to Hawai'i in the middle of the pandemic. Upon arrival, we met an entire battalion of unsmiling National Guard troops scrutinizing vaccination credentials before admitting us tourists into their tropical paradise.

Hawai'i started out as a paradise for birds, too, but now the archipelago offers a tragic glimpse into the future of what the edge of a bird-less dystopia might look like. Formed 30 million years ago out of molten rock from volcanoes near the center of the Pacific Ocean, these islands offered a pristine tropical climate, largely free of predators and disease. Its birds evolved into a spectacle of colors, some with beaks that curve in half circles and others with drill-like pinchers reminiscent of a dentist's tools. But once outsiders arrived—first from Polynesia and then Europe—hordes of rats, cats, fire ants, mosquitoes, mongoose and feral pigs, goats, sheep, and cows came along with them. These invaders fed on birds' eggs and sometimes the birds themselves. Gradually much of Hawai'i's rainforest was trampled under their hooves. While the decline of a species on the mainland can take decades, Hawai'i's extinctions are happening from one year to the next.

"Sometimes I feel like my job is like a hospice nurse," says Justin Hite, a field supervisor who works with forest birds deep in Kaua'i's rainforest.

The work to save birds in North America has mostly relied on advances in research, saving habitat, and rebuilding individual species one at a time. But what happens when that toolkit isn't enough? Hawai'i's nontraditional approaches could very well be the answer. There also are futuristic endeavors brewing on the mainland in laboratories, zoos, and genomic research centers. The San Diego Zoo Wildlife Alliance has collected 10,000 bits of live tissue from threatened animals to prepare for a day when that's the best way to save birds and other wildlife—a genomic insurance policy of sorts. A team of researchers led by a scientist in the mountains of North Carolina has gone a step further: trying to resurrect the once abundant Passenger Pigeon, declared extinct back in 1914.

The situation in Hawai'i is dire, but is it hopeless? It's not hard to find people in the bird sciences who think so, and given the needs of other species throughout the hemisphere, they say the money going to these islands is better spent elsewhere. "That ecosystem is not recoverable," says Paul Schmidt, a former top official with the U.S. Fish and Wildlife Service. "Humans have devastated that ecosystem. You can't put that genie back in the bottle."

In fact, the United States has never made native Hawaiian birds the priority given what's at stake here. One funding assessment found that although a full third of all the nation's officially listed endangered species live in Hawai'i, only about 4 percent of funding spent under the Endangered Species Act goes to the islands. Hawai'i is often overlooked by guidebooks, research conferences, and bird organizations, in part because it's so different from the mainland and partly because making conservation progress here is such an uphill slog. "I think of Hawai'i as a cautionary tale," says Noah Greenwald, an endangered species specialist with the Center for Biological Diversity based in Tucson, Arizona.

Despite the odds, or maybe because of them, Hawai'i is now forced to push the boundaries of conservation. A new generation of researchers is taking fresh approaches to problems that have plagued the state for decades. André and Helen Raine, two world-traveled scientists who've settled on Kaua'i, launched their own company to find solutions before it's too late. They're slowly reversing the collapse of the seabirds that looked like lost causes a few years ago. Far up on the giant volcanic mountain of Mauna Kea on the Big Island, a restoration project in one of the state's most productive rainforests is bringing the landscape back to health. Now the refuge is home to an exotic group of native forest birds that might otherwise be extinct. The biggest gamble is the *Wolbachia* mosquito project, the largest and most expensive mission of its kind in the world. "Hawai'i is a small place, and the problems are so severe that everybody works together," says Chris Farmer, head of Hawai'i programs for the American Bird Conservancy. "We have a limited amount of time. We're in the stage where all options are on the table."

Like a Zombie Movie

The first time André Raine dropped from a helicopter to an otherwise unreachable colony of burrowing seabirds on the northwest edge of Kaua'i, he was startled by the devastation. "You come to this ridge where nobody had been for possibly forever," he says. "The first time I went there it was amazing. There were so many dead birds because of the cats, rats, pigs, and barn owls." He found three burrows attacked in a single ten-foot area. A rat had killed a chick in the first burrow, a cat ate an adult bird in the second, and a feral pig destroyed the third burrow. "It was like we were seeing a microcosm of the whole issue for them in the mountains," Raine says. "We've got to get protection for these birds."

André and Helen moved to Hawai'i twelve years ago with their two young children to find a more peaceful place to live. They first worked together in Zambia, living in a pup tent for months at a time. They went on to South America and then to Europe, where at times they tangled with violent poachers. On the Mediterranean island of Malta, Helen once caught a poacher on video, which led to the man's arrest. When he was released, he drove to the Raines' home and confronted André while Helen, her mother-in-law, and the children looked on. "He had me up against the wall, and he's yelling at me, 'You're the one that filmed me.'" Meanwhile, Helen ran inside, grabbed their camera and rushed to the top of the house. "I was like, 'That's right, and she's filming you right now.' He looked up and saw Helen on the roof with the video camera." Then the guy jumped in his car and left.

Some years later when André applied for the director post at the Kaua'i Endangered Seabird Recovery Project, he was asked if he could handle any tension that came with the job. "They asked him how he was with confrontations and difficult situations," says Helen. "He said, 'Yeah, let me tell you about that.'"

Yet Hawai'i presented an entirely different kind of friction. Helen went to work in conservation with wetland birds, and André began to look for strategies to help two seabirds, the Hawaiian Petrel and Newell's Shearwater. Both birds build burrows on the peaks and cliffs far up in the mountains, and both were protected under the Endangered Species Act early on. These birds spend most of their time at sea. The petrel, a dark gray-brown bird with a white face and belly, is named 'ua'u by native Hawaiians for its haunting nocturnal call. It's one of the ocean's most wide-ranging marine species, flying as far afield as Alaska and California to search for food for its family. The Hawaiian name for the Newell's Shearwater is 'a'o for the donkey's-bray-like call it makes in its burrow.

Both birds are shaped like tiny, winged dolphins and are built to withstand life on the high seas with waterproof feathers and the ability to dive deep into the ocean to fish. They play vital roles circulating nutrients, consuming and being consumed, and living more than twenty years. The seabirds come to Kaua'i and other islands to breed and raise their young in the spring and fall, but the islands in Hawai'i are not a hospitable maternity ward. Powerlines, virtually invisible after dark, crisscross their flight paths on Kaua'i. Lights from coastal development, hotels, and Friday night football stadiums confuse the birds because they navigate by the moon. One detailed study found the Hawaiian Petrel had lost 78 percent and the Shearwaters 94 percent of their populations over the previous two decades.

Raine can't shake the image of the predators that are constantly going after the bird colonies. "I'm a fan of zombie movies," he says. "That never-ending remorseless horde is exactly what it's like up there. It really is. Each season, you're like, 'Here we go again. Here they come.' And they're scratching at the gates."

Raine is a methodical and relentless scientist. When he first arrived, he'd often pack supplies and a sleeping bag and head up into the mountains for days at a time to study the seabird colonies. Lean and stoic, Raine admits the work is so physically tough his joints are starting to give out. "His knees are bad, but he just keeps going," says the American Bird Conservancy's Brad Keitt, who works with Raine. These trips into the interior can be especially treacherous. Raine took us up the mountain during a rain shower that turned the steep paths into slippery mudslides. He offers a simple warning as we scramble along a path just a few feet from a cliff that drops thousands of feet straight down to the Pacific. "If you're going to fall, be sure to fall to the right," he says. "If you fall to the left, it's all over."

Once Raine had sized up the seven seabird colonies he works on scattered around the island, he began documenting with audio recorders the attacks on birds. He attached tags that worked like grocery checkout readers to record the comings and goings of birds from their burrows in order to count them. Next he and his staff needed to figure out how often birds collided with the powerlines that encircle Kaua'i. The lines were taking a toll on the seabird populations, and the situation had caused a long-running dispute between environmental groups and the Kaua'i Island Utility Cooperative, the company responsible for the lines. Environmentalists finally sued the utility, leading to an agreement in 2010 to address the problem. Even then, the company was slow in responding, insisting the collisions were rare, says Keitt, director of oceans and islands programs for the American Bird Conservancy. "The utility company was extremely obstructionist," he says. "It has taken a ridiculously long time."

Raine's staff spent hours sitting under power lines at night, watching for collisions and hearing the haunting twang of birds hitting the wires. "We thought since it makes a really audible sound, that was something we could record," he says. "Then we might be able to actually understand the scale of the problem." An earlier federal analysis had reported 1,800 annual deaths, but after Raine posted recorders along the base of the lines, he documented a stunning rate of collisions nine times higher than the federal estimate: Birds hit the wires nearly 16,000 times a year, which they found killed almost 30 percent of the birds. When Raine's staff fitted seabirds with trackers to map the flight paths, they were able to pinpoint where most collisions occurred. The maps and recorders helped forge agreements with the utility on fixes, such as lighting the wires, using deflectors to scare the birds off, and in some cases moving the lines underground. "It's been a long battle," says Raine, "but now we're actually in a good spot because everybody understands the problem. It's really bad, but you can deal with it."

The results have been dramatic. The overall percentage of bird calls in the breeding grounds is increasing by about 10 percent each year, which gives a rough sense of the rising population. "It's really fantastic," says Raine. "It shows that the management is really working." Once the power company was held liable for collisions under the Endangered Species Act, it had to pay millions of dollars in remediation that funds safeguards for the seabirds. The colonies are cordoned off with heavy fencing to keep out the pigs. They've set up traps that can kill up to two dozen rats in succession. They've started shooting the feral cats and Barn Owls that proved to be the seabirds' chief predators. As a result, the number of new chicks hatching in the colonies is gradually growing.

The Raines are now experimenting with a new approach to their work. André left his job with the Kauaʻi seabird project, a joint state and university agency, and he and Helen launched their own firm, which is carrying on the seabird work as an independent contractor. Their company, ARC, for Archipelago Research and Conservation, operates out of the second floor of a former hotel in the artsy coastal town of Hanapepe. They've left behind the government approvals that slowed the pace of decisions. "It's quite liberating just to be able to say, 'Okay, we're going to do this project,' and just do it without all the bureaucracy." The back stairs creak with the coming and going of fourteen recent graduates and interns they've hired. Muddy boots, taxidermied birds, and seabird skulls litter their offices. In one room, a half dozen staffers lean over their computers, analyzing the findings from recordings, tracking devices, and the powerline data. "There's a risk, but we're in control of our own destiny," says André Raine. "Hawaiʻi is a place where you've got to be creative."

Reviving Hawaiʻi's Behemoth

Depending on how you measure it, Mauna Kea on the Big Island of Hawaiʻi is the world's tallest mountain, reaching 32,696 feet—about 3,660 feet taller than Mount Everest. The catch is that Mauna Kea is an inactive volcano, and two thirds of it is hidden beneath the surface of the Pacific Ocean. As a rule, mountains are measured from sea level, which gives Everest its edge. But any way you look at it, Mauna Kea is a formidable force in Hawaiʻi, its slope so gradual you can travel nearly halfway up the mountain by the island's main highway, then slowly make your way to above 10,000 feet via the bumpiest of gravel roads. For all its monstrous size, though, Mauna Kea, which means White Mountain in Hawaiian, is a shell of its former self. Much of the top of the mountain is a moonscape of barren rock, leftover lava, ash, and gravel. All along its sides, the remnants of decades of cattle grazing dating back to the early 1800s have left the forests and fields foraged into oblivion. Feral descendants of those pigs, sheep, and cattle still roam the mountain and keep much of the growth from regaining a foothold. A mountain that is a symbol of Hawai'i from a distance is a battered beast close-up. "This is one of the last big native dry forests left in the world. It's incredible," says Chris Farmer, the American Bird Conservancy scientist. "It's also tragic."

Two notable efforts are in the works to restore portions of Mauna Kea to support the native birds that live here. One is the Hakalau Forest National Wildlife Refuge on the eastern slope, home to six native forest birds. The other is the work along the southwestern slope to protect the last Palila honeycreepers found nowhere else. Hakalau has been a surprising success, showing that it is possible to rebuild the Hawaiian rainforest. The Palila rescue is far slower, since replanted trees can take two decades to grow on the dry side of the mountain. But a donut-shaped ring of the bird's range around the mountain from 7,000 to 9,000 feet is slowly filling in as a home for the 1,000 remaining Palila honeycreepers. "It used to be that you could walk up here and see almost all the way down the mountain," Farmer says. "Now there's a lot of understory, so the mountain is slowly recovering."

The Palila is a bright yellow, gray, and white bird with a sweet, staccato whistle of a song that sometimes seems to end in a question. Palilas once extended from almost one end of Hawai'i's largest island to the other, but now are limited to Mauna Kea. The Palila relies on the Māmane tree, eating practically every part, including seeds, flowers, and leaves that haven't even had time to fully mature. Farmer drives us up the face of Mauna Kea in his four-wheel-drive truck to see the mountain and try to find the Palila. He has worked on protecting this bird for thirteen years, first with the U.S. Geological Survey and then with the Bird Conservancy. From the moment we enter the bird's range, Farmer starts catching whiffs of its songs that we ourselves can't hear. But as if to demonstrate the bird's fragile status, we spend almost the whole day without spotting one in the very heart of its territory.

"Realistically, one fire, one hurricane and we lose the species," Farmer says.

We finally give up and head back down the mountain, stopping for a final check on the way. We unlock the gate of one last preserve and start walking through the grasslands. Suddenly, Farmer stops, tilts his head, and holds up his hand. "He's close," he says and begins running through one field and into the next, climbing over the lava rocks. More than once we tumble to the ground. Finally the Palila pops up right in front of us, feasting on the bright yellow flower clusters in a low-slung Māmane tree. The day's last direct sunlight turns the bird's yellow head to a golden hue, and we stand there silently, taking it in.

Once the bird flies off, Farmer starts talking about what it's like working with birds that can disappear at any point. "Anyone who's been here for any length of time has dealt with birds that have gone extinct. You hear the horrifying tales: 'This was the last bird. I heard it. I couldn't see it. It was never seen again,'" he says, recounting the stories he's heard of the 2004 loss of the Poʻouli, one of the most recent Hawaiian birds to go extinct. "A lot of this seems melodramatic, but extinction is always there. It's a reminder that you're—you're, I don't know what the words are." He stops and then adds: "If you have enough habitat to protect a Palila, the forest is in good shape. Everything else is going to benefit if you save the birds."

The other restoration project on Mauna Kea, Hakalau Preserve, is almost halfway around the mountain. We make the trip the next day with retired biologist Jack Jeffrey, who now works as a guide for what's become one of Hawaiʻi's most important replenished rainforests. Like the Palila's territory, much of the Hakalau forest was destroyed over the years by the feral animals. But when the state did a survey in the late 1970s of where the most native birds were on the Big Island, the results pointed to this chunk of land that still held some of its original native trees. A decade later, the U.S. Fish and Wildlife Service started replacing invasive plants and trees with native ones, building fences, and ridding the area of wild pigs and cattle. Jeffrey helped oversee the reconstruction as the lead biologist, and on weekends he would recruit volunteers to help with the digging, lifting, and planting.

One day in the early 1990s while the project was still in the initial stages, Jeffrey got a call from a University of Hawai'i professor who'd heard about the project. "I'd like to bring my environmental law class to the refuge for the weekend," the professor, Denise Antolini, told him. "Can we do that?" Jeffrey says he was glad to have energetic students help out with the exhausting, sweaty work. One morning Jeffrey sat down with Antolini on the front steps of the volunteers' cabin, looking out over the fields. "It was just open pastures. There were no trees, nothing," he says. "You can see the sun rising out of the ocean. And I said, 'Someday, this cabin is going to be surrounded by trees. It's going to be in the forest, this cabin. All the birds are going to be here. But it's not going to happen in my lifetime. We all know how long it takes a forest to grow.'"

Antolini brought her classes back almost every year so her students would gain an appreciation for the labor of conservation. Jeffrey eventually retired in 2008 and started his work as a guide and photographer. Several years ago Antolini gave Jeffrey a call ahead of her next visit and invited him to join the group. "We haven't seen you in a long time," she said. Midway through the weekend, he and Antolini ended up at that same cabin one morning. "And she said, 'Do you remember what you said the first time around? That this cabin was going to be in the forest and the birds would all be here?'" She looked around and marveled at the transformation with flowering trees that now reach fifty feet high and carpets of delicate ferns on the forest floor. "'Well, you can't even see the ocean anymore because all these damn trees are in the way.'" Jeffrey almost never stops joking, but he suddenly turns serious as he contemplates this story. "It was an aha moment," he says.

Mauna Kea is not an easy mountain to restore. Its sheer size—millions of acres along its slopes—makes the recovery work laborious.

Disputes and lawsuits over the early pace of work delayed progress, but Farmer says the many agencies with a stake in the mountain are now working together, and he sees progress with every visit to the Palila territory.

The best example of what can be accomplished is on the Hakalau preserve. Six species of endangered birds forage in the now mature koa and ʻōhiʻa trees. One of Hawaiʻi's most recognizable birds—the ʻIʻiwi honeycreeper with its brilliant scarlet plumage and a beak that curls into a half circle—moves slowly among the fluffy red blossoms, called lehua flowers, of the ʻōhiʻa trees. Sturdy fences now encircle the core of the 33,000-acre preserve to keep out four-legged intruders. When you pass from the old grazing pastures of flattened grasslands into the restored refuge, it's like stepping into a technicolor world for the first time—one that includes a hopeful soundtrack: You can hear these endangered species singing as soon as you walk down the main path into the preserve.

Mosquito Birth Control

The realization that mosquitoes were killing off Hawaiʻi's forest birds came a little at a time. Species with melodious names such as ʻAkikiki and ʻAkekeʻe went missing like victims in a murder mystery. Ten native species, most of them the island's unique honeycreepers, went extinct by the mid-1900s in the lowlands where most people live. The native birds seemed able to survive only in the island's higher, cooler elevations.

Finally, in the mid-1960s, scientists brought some of the birds down from the mountains and put them in cages in the lowland, which soon solved the riddle. The birds were falling victim to a strain of avian malaria traced to mosquitoes whose range was limited to the warm lower elevations. "If you read the old naturalists' accounts, the lowland forests were just full of birds . . . beautiful honeycreepers—red birds, green birds, yellow birds," says Dennis LaPointe, the scientist at the center of

Hawai'i's mosquito eradication project. "So we had a total loss of these native birds in the lower elevations. I noticed it immediately when I moved here. I thought, 'This is going to take some getting used to.'"

Then a decade ago, the U.S. Geological Survey noticed that the island's climate was shifting enough even in the higher elevations to undo the delicate balance for birds. Higher temperatures were not only causing seawater levels to rise, but also sending warmer air both inland and higher than ever. Mosquitoes buzzed steadily upward, too. At first the warming was slight, but in the last few years, temperatures started increasing faster than anyone expected. Waves of mosquitoes showed up in higher elevations in surprisingly large numbers in 2020, and then field-workers started having trouble even locating some of the native species. "Now we're finding mosquitoes at the very center of the forest," says LaPointe. "It's a pretty grim picture."

It's been almost a decade since LaPointe and representatives of twenty different nonprofit and government agencies first started preparing to use *Wolbachia* to halt the further spread of mosquitoes. *Wolbachia* has been used to prevent the spread of human disease carried by mosquitoes in smaller projects in Texas, California, Australia, and Singapore. The question in Hawai'i is whether it will work on a far broader scale throughout the interior of the islands.

Lisa "Cali" Crampton, director of the Kaua'i Forest Bird Recovery Project, watches over the birds most in danger on her island. Sitting on her open back porch overlooking the azure-blue Pacific Ocean on the southern coast of Kaua'i, she explains how she was drawn to the islands twelve years ago by the immense beauty and the chance to work with Hawai'i's famous collection of native birds. But from those very first days, she found herself battling the invasive southern house mosquito, or in Latin, *Culex quinquefasciatus*. "I'm more worried than I've ever been," Crampton says. "I just hope we can get the mosquitoes here in time."

The phone rings in the early afternoon of our visit, and Crampton rushes off to have one more conversation about whether it's possible to move more quickly on the rescue mission. When she returns, she picks up where we left off, but her mind is clearly on her frustrations, the weight of these species resting on her shoulders. A few minutes later, as we talk about the friction that comes with her job, she stops and breaks into tears. As she recovers her composure, she talks about how quickly things are deteriorating. "I asked my boss the other day. I said, 'Can I take my son into the field so he can see the 'Akikiki before they're gone?'" she says. "We've got to help them. They're fighting tooth and nail to the very end. So we're fighting tooth and nail to the very end."

The mostly young biologists who track the birds up in the mountains find it painful to watch species like the 'Akikiki evaporate in front of them—eating away at Hawai'i's natural foundation that feeds the vegetation, maintains the rainforest, and helps keep the island's wildlife balance in check. Avian malaria is an ugly death. First the mosquito bites the bird on the bare skin of its eyes or legs. It deposits a parasite that ruptures red blood cells, causing internal organs to swell. Eventually infected birds become lethargic, and many will die from fever.

"My first year out, I found like fifteen nests," says Justin Hite, the field supervisor for the recovery project, who's worked in Hawai'i for the last eight years. "This year, I've found zero and my crew has just found three. So there's just three breeding pairs at our primary site, and we think one female just died. The next thing we know, the male is flying around looking for her and he can't find her." Hite is an outgoing, upbeat staffer, a veteran of projects like this in fifteen countries, but he can't hide his frustration. "My God, if we're not going to learn from this, then shame on us. We may need to say goodbye to the 'Akikiki," he says. "We may be too late for the 'Akikiki, but we're not too late for the other birds."

Preparing for the mosquito project is like getting ready to launch a moonshot. The costs increase up to $10 to $12 million a year once the releases begin. The details must be approved by Hawai'i's many layers of bureaucracy, including state and federal agencies, health departments, and individual islands' bird recovery groups. The moonshot analogy is apt in other ways: Nobody can say for sure how well—and even if—the *Wolbachia* mosquitoes will stop the spread of malaria. "If you're an endangered forest bird on the brink of extinction, it just takes one bite and you're gone," Crampton says. If malaria spreads, about a dozen species in the Hawaiian honeycreeper family are expected to die off—a startling loss even by Hawai'i standards.

"It's our best hope," says Teya Penniman, who was the project's coordinator during its first years of preparation. "It's a grand experiment, and I think that the scale and the challenge of the project, and the fact that people are taking it on, speaks to the commitment and the passion and the hope we can make a difference."

Hawai'i's mosquito project is not just a test of the birth control concept. It will also gauge public willingness to accept an experimental treatment spread throughout a wide landscape of rainforest. The project staff is working to publicize how the release will help and has come up with a pithy slogan—"Birds, not mosquitoes"—to drive home the goal. So far, the reaction has been positive, including among native Hawaiians who've traditionally been skeptical of government experiments.

Sabra Kauka, a native Hawaiian who's served on local boards, in government and as a teacher, said the health of indigenous birds is important to the island's original inhabitants, who've long seen birds as a part of their culture. Kauka has gotten to know the island's conservation staffers during ceremonies in which she blesses injured seabirds that have recovered and are being released back into the wild. She sees the mosquito plan as a long shot. "You do want to hear the birds in the forest forever. But the reality is, the temperatures are rising, and the mosquitoes are going higher and higher," she says. "I have a great deal of hope, but it's tempered with reality."

The final year of preparations has been devoted to understanding how many mosquitoes have spread through the highest elevations and exactly where they are. The field crews scattered traps throughout the mountains to take samples that will dictate how many and where to release the *Wolbachia* insects. LaPointe, the chief mosquito scientist, said the team has done enough preparation to believe there's a good chance of rescuing an entire suite of endangered birds. "When I'm being optimistic, I would say it's an 80 to 85 percent chance of success. Then I can sit down and think of all the things that could go wrong and my confidence drops," he says. "You know, until we attempt it, we're never going to know. If we pull it off, we'll achieve something great."

LaPointe voices one caution: This may not be a permanent fix, he says, since the landscape in Hawai'i is constantly changing. He hopes the treatment will keep the forest birds alive until something more lasting comes along. Scientists have started breeding the most threatened honeycreepers in captivity for the day they may no longer exist in the wild, and they've begun talking about other remedies expected to develop in the coming years. "In a decade or two, maybe we can get to the point where we have a genetic solution," he says. His view is echoed by many of the scientists in Hawai'i and elsewhere: Conservation through genetics—analyzing genes for disease, expanding biodiversity for endangered species, even cloning animals that are down to a few last individuals—is going to be the next development in the conservation toolbox.

The science of genetic rescue has taken off in scattered research centers, zoos, and universities in preparation for a time when the genes of species like these honeycreepers can be programmed to be immune to malaria. How far off is that day? The best place to find an answer is at the San Diego Zoo Wildlife Alliance in Southern California.

Conservation Genomics

San Diego, California

At first glance, the six giant steel tanks in a corner office of the San Diego Zoo look like they belong in a microbrewery. But when the staff pops open one of the lids, out comes a waft of frigid vapor and rows of colorful vials holding precious samples of cells. This is the Frozen Zoo, a zoo within the zoo, where life of all kinds is on permanent standby at minus 320 degrees Fahrenheit. That's the temperature at which life can be preserved and stored indefinitely, which is the purpose of this singular project at the San Diego Zoo Wildlife Alliance. For four decades, the Frozen Zoo has collected cultures, sperm, and embryos from endangered apes, birds, rhinos, sloths, reptiles, and now plants. When the originator, Kurt Benirschke, started the project in 1975, he put a poster on the wall that's still hanging today. "You must collect things for reasons you don't yet understand," it reads.

The value of this particular collection is no longer a mystery. The vials, 10,000 of them from 1,200 species in all, contain the building blocks for a new way of helping endangered species survive. Think of it as conservation through genomics, where the focus is on the inside of the animal. Through billions of strands of genetic material, scientists are researching how to strengthen, repair, and even rebuild species. It's also possible that genomics can eventually help wildlife withstand the disruptions of changing climates that are altering habitats faster than species can adjust. "I imagine a future in which there's been tons of technological, ecological, and ethical advances, and we might be able to use the tools of gene editing or genetic engineering to try to help species adapt to the changing environment," says Beth Shapiro, an evolutionary molecular biologist at the University of California at Santa Cruz and author of the book *How to Clone a Mammoth: The Science of De-Extinction.*

It has taken years to perfect the mechanics of collecting and storing these cultures in the thousands of vials holding living cells. Early on, when there was just one tank, the offices once lost power and the entire collection was ruined. Now it's backed up with generators, duplicated with an entire replica of the collection offsite, and monitored by a network of cameras and alarms that not infrequently roust Marlys

Houck, director of the Frozen Zoo, out of bed to make sure things are okay. "It's usually a false alarm," she says. There's a lot at stake in keeping the collection safe; it's the only one of its kind in the United States. "I don't believe there's any place like this in the world," says Cynthia Steiner, associate director of genomic conservation at the Wildlife Alliance.

Advances have come rapidly in the past couple of years. In 2020, a California-based nonprofit called Revive & Restore used cells from the Frozen Zoo to clone the first endangered species for conservation purposes, a short, stocky Przewalski's horse from Mongolia named Kurt that had gone extinct in the wild but was reintroduced through captive breeding. Then in 2021, Revive & Restore, along with a pet cloning firm and the U.S. Fish and Wildlife Service, cloned an endangered black-footed ferret named Elizabeth Ann, using DNA from a ferret that had died thirty years earlier and was stored in the Frozen Zoo. This was the first rebirth of an endangered species in the nation with its conservation as a goal, and it was especially significant since the ingredient cells had been stored for decades.

One of the early genomic achievements was solving the problem of a recessive gene in the California Condor, a major recovery target for a consortium of agencies in the West including the alliance. The condor went extinct in the wild for a time, but has slowly built its population through captive breeding to about five hundred birds, most of them in the wild in parts of California, Arizona, and Mexico. Researchers found an approximate location for the lethal gene that prevented some of the birds from reproducing. This enabled them to identify which birds had the gene and avoid mating two disease carriers. The goal is to have the ability to edit a species' genes to repair disease. "But that's something we don't have yet," Steiner says.

As chief conservation officer for the Wildlife Alliance, Nadine Lamberski looks at the Frozen Zoo and genomic research as a backup plan for when things go badly wrong. "We hope we don't have to use them. But we can't be so idealistic that we don't have a backup plan," she says. The potential in genomics is significant, and although the alliance is putting the Frozen Zoo's library of cells to work, Lamberski says there's still much to be discovered about its uses. "I'm not sure we know yet all the things that are going to branch out from this technology."

The next challenge is whether, or perhaps when, it will be possible to bring back an extinct species. Each of these research steps moves that day closer, but there are still barriers ahead. The Revive & Restore nonprofit is betting it's a matter of years, not decades. The nonprofit has already chosen the bird it hopes to bring back. It started raising the millions of dollars it will take and has found an obsessive scientist who's devoted his life to the new discipline of de-extinction. "We really want to figure out how to literally crack this egg," said Ryan Phelan, Revive & Restore's cofounder and executive director.

Putting De-extinction into Practice

Brevard, North Carolina

Ben Novak steps through the front door and begins a tour of the laboratory where he plans to rebuild the Passenger Pigeon, which has been extinct for the past one hundred years. "This first room is where we'll incubate and work with eggs," he says. "And here's where the birds will be bred." As we walk a little farther, he moves into a rapid-fire explanation of the steps he sees ahead: incubators will go in one corner; pigeons that act as foster parents will sit here; with luck it will take about three weeks for the eggs to mature. Then he comes to the point he's been thinking about for decades. "I have every intention," he says, "that the world's first Passenger Pigeons will be born right here."

The laboratory we're viewing doesn't actually exist yet. We're strolling along the driveway of Novak's house in western North Carolina as he paints a picture so vivid he seems to see the birds already in their places. The idea of bringing back the Passenger Pigeon first grabbed him when he was a thirteen-year-old high school student in North Dakota, and never let go. There was something about this bird, he says, that fascinated him the first time he saw a photo of its red eyes, robin-like rouge breast, iridescent feathers, and bright-red feet. For most of the next twenty years, his overflowing enthusiasm was all that propelled him in this solitary pursuit, one so far ahead of its time there was no name for the concept. Just when Novak thought he might give up, three things happened in succession: He discovered he wasn't alone in his goal; genomic technology began to advance; and the field found a name that got people's attention: de-extinction.

Novak, in his mid-thirties, is now the chief scientist at Revive & Restore, launched by Ryan Phelan and her husband, futurist Stewart Brand. A little more than a year ago, Novak and the nonprofit lined up a $5 million gift to confront the scientific barriers that stand in their way in the Passenger Pigeon project. The funds helped them organize a worldwide network of forty genomic laboratories that will take on parts of the experiments that make the cloning of birds an especially difficult challenge compared to the horse and the ferret successes. "This isn't just about Passenger Pigeons," says Novak. "This is a shared barrier for cloning all birds. This should in the next three to five years get us to the point where we could really, actually start creating Passenger Pigeons."

"Are You Ready?"

Stewart Brand, the ardent environmentalist best known as the creator of the countercultural magazine the *Whole Earth Catalog*, introduced the idea in 2013 in a TED Talk titled, "The Dawn of De-extinction: Are You Ready?" Wandering a stage in Long Beach, California, Brand began and ended with the story of the Passenger Pigeon. Up until that day, the idea of reviving extinct species had been a vague and theoretical concept. But Brand's methodical talk on bringing back extinct species firmly attached the quest to the Passenger Pigeon, a bird native to the United States with a history that made it one of the most consequential species on the continent. "This had been the most abundant bird in the world that had been in North America for six million years. Suddenly it wasn't here anymore," said Brand, lean and balding, with a strong, confident voice from his years on stages like this one. "What happened?"

Then he told the story of the pigeon's rapid decline from as many as five billion birds that would flock together in great roiling clouds as they traveled from one roost to the next. That made them vulnerable to trapping and mass hunting, and people killed hundreds, sometimes thousands at a time. The Passenger Pigeon was a popular source of food as well as a target for sport hunting. By the time people realized the entire species was at risk and banned hunting in several states, it was too late. The last Passenger Pigeon, a bird named Martha, died at the age of twenty-nine in the Cincinnati Zoo on September 1, 1914.

Brand and his wife started Revive & Restore to begin the work of bringing back such species as the woolly mammoth, the Carolina Parakeet, the Eskimo Curlew, perhaps even the Ivory-billed Woodpecker. "I think it's time for the subject to go public," he said, and addressed his TED audience. "What do people think about it? Do you want extinct species back? Who wants these species back?" A round of applause rippled through the theater.

The reception wasn't so warm as the news of the plans spread in newspapers, magazines, and eventually dozens of scientific journals and books. Many of the articles referred to the movie *Jurassic Park*, even though the DNA of dinosaurs is known to be too ancient to reconstruct. Plenty of ethical questions got batted around, too. What's the point of reviving species if there's no longer sufficient habitat? Would bringing a species back disrupt the balance of nature that's evolved since it disappeared? Beth Shapiro, the molecular biologist who's been a leader in the field and is now on the Revive & Restore board, says workable answers to the habitat questions are essential, but she says many of the criticisms are without merit. The part of the clamor that bothered Shapiro most was how the debate got so far ahead of the science itself. "The problem is that we're being condemned for doing things we can't even do," she says. "People get excited about science fiction. They think that scientists are doing things when we're just thinking about it, and trying to imagine how we could do this, trying to bring solutions to these amazing problems, bringing as many tools as possible."

The steps for bringing back the Passenger Pigeon each carry their own complications. In short, researchers must decipher the genome of the Passenger Pigeon that provides the biological blueprint of what made this bird the way it was. Then they must choose its closest relative, likely the Band-tailed Pigeon, common in the West, and edit out the genomic differences between the two. This would be done with a gene-editing tool that works like a pair of molecular scissors called CRISPR, a discovery that won the Nobel Prize in 2020. Then they must breed the new species they've created in captivity. Each step along the way is complex, which is why Revive & Restore set up the network of international researchers to take apart the processes and figure out how each applies to birds.

The complexity of this work is best illustrated by what it takes to map the billions of particles of the genome of both the Passenger Pigeon and the Band-tailed Pigeon. Here's how Novak described the mapping challenge now completed by his team and other labs: It starts with multiple samples of tissue taken from toe pads and bones of the birds. Then they break those DNA strands apart so the sections are small enough to match up like a puzzle. "So, you've blasted them into tens of thousands of smaller pieces that you can sequence," he says. "The problem is that you don't know how the puzzle went together before you blasted it apart. Now you have to put it back together. So, it's literally as if you were to take a 1,000-piece jigsaw puzzle that's all one color and then have someone cut all the pieces up, like throwing them in a shredder, then hand it back to you and say, 'Give me the puzzle.'"

Once the genome is deciphered, or sequenced, it can serve as the roadmap for reshaping the genes of the Band-tailed Pigeons to be as much like the Passenger Pigeon as possible. That means that the cloned species isn't actually a Passenger Pigeon. It would be more like a replica with a few minor differences, perhaps in its eye color or plumage. As Stewart Brand put it, "The result won't be perfect. But it should be perfect enough. Nature doesn't do perfect either." The final hurdle—and it's a big one—is how to reintroduce a cloned Passenger Pigeon into a world it left more than a century ago. Would it have the instincts of the original pigeons? Would there be a place for these birds? What's the purpose of bringing this bird back to life? Ben Novak has been thinking about these questions for decades as well.

Giant Aviaries Full of Pigeons

Once we complete the tour of Novak's future laboratory, we move to the wooded backyard of his home. "You come out somewhere like this," he explains, "and basically make a football-field-size aviary where all your baby pigeons, after they're raised, get put in here." If all goes as planned, the hatchery and aviaries will begin producing first hundreds, then thousands of pigeons until their captive populations might reach 10,000 birds ready to be released into the wild. This is where Novak makes the case for how Passenger Pigeons can help restore the environment of the country's eastern woodlands.

Before their extinction, the pigeons traveled in huge flocks, roosted en masse in the forest to feed on acorns, beechnuts, and chestnuts in the fall and berries, flowers, caterpillars, and insects in the spring. Their numbers were so great, they'd break branches, pummel the ground, and leave behind mounds of pigeon poop. This served to rejuvenate the forests in the same way that fires do. This benefit is a central part of the argument for bringing Passenger Pigeons back. At a time when fire is often suppressed and many forests have regrown without much variation in tree species, the Passenger Pigeon's revival could help diversify the nation's forestlands. "They're not just a significant player in the forest," Novak says. "They're the most significant player that the forest has ever had."

Novak's now working on a demonstration to prove the theory. He plans to re-create the impact of massive groups of pigeons roosting on a track of land by carting in truckloads of pigeon poop, digging up the ground, and using chainsaws to break down trees to imitate the impact of thousands of roosting birds. Then he'll measure how the forest responds. The prospect of this experiment does beg a question: If flocks of Passenger Pigeons are so destructive, how will the public react to the ravages these birds would create? When Novak gets wound up, his voice rises a few pitches and his words tumble out in a rush. This question sets him off.

"So the choice is to lose everything, invest in a bunch of management that won't scale up—or put up with some bird shit. Like, wash it off your car. Get over it," he begins. "If your viewpoint is that a billion Passenger Pigeons might be annoying, then you don't care about nature. And you can't profess to care about endangered species, if what you're saying is you want the natural world to be in a little box that you visit with your visitor's pass on weekends and never touch your house, or your yard, or your life . . . that's not going to work. It's either a future where we pave the planet and make it all farm fields and we let everything else die and live in zoos, or we learn how to integrate human lifestyles and structure into a world that is wild. And I think Passenger Pigeons are a huge leap toward getting people to think about living in a wild world in a modern context."

By the end of 2022, the construction of the pigeon laboratory is finishing up in Novak's driveway, and a half dozen separate research projects are under way from Revive & Restore's international lab network. The timeline now calls for hatching the first birds in five to seven years, and releasing flocks in perhaps another decade. If all goes according to plan, the genomic tools will develop alongside the de-extinction research and play a steadily more important part of strengthening species. "I think in ten years' time, these tools could become a huge part of the conservation toolbox," says Phelan. "That doesn't mean that de-extinction will be ubiquitous. But I am hoping that extinction will start to become something that happened a century ago."

The Greater Sage-Grouse, with a population spread across eleven Western states, sits at the center of a decades-old struggle pitting the health of the species against the regional economy. The males, adorned with what looks like a fur stole, put on a riotous mating performance in the spring.

9

COEXISTING WITH THE BIRDS

Flint Hills, Kansas

The day of his high school graduation, Josh Hoy tossed his saddle, an old army sleeping bag, and a few clothes into the back of his silver Nissan Stanza station wagon and headed west to look for work as a cowhand. He had plenty of experience—and it started at an early age. One of his first memories was of getting up at 3:30 a.m. for a cattle roundup on his grandparents' ranch, the Flying H, near Cassoday, Kansas. A handful of old-time cowboys gathered in the kitchen telling stories over his grandmother's breakfast of chicken fried steak, hash browns, and biscuits. Then they saddled up and rode off together—with Hoy on his own horse. "I was probably six or seven years old," he says. "That made a big impression on me." By twelve, he was working on ranches around his home in the Flint Hills of southeastern Kansas. After he got a driver's license, he started taking ranching jobs around the state, spending all the time he could outdoors. "The natural world was just everything that mattered to me," he says. So when he was done with the classes that never held his interest, he knew what he wanted to do.

For the next ten years, Hoy traveled to fourteen states in the Midwest and West, working as a hired hand, usually for $40 a day. Home became his station wagon. "I lived in that thing for years," he says. Despite his leather chaps and wide-brimmed hat, Hoy never looked the part of a cowboy. He was on the chubby side, and he'd always looked so awkward on a horse that some ranch managers took one glance and shook their heads. But Hoy was a self-starter, a quick study who paid close attention to details. In time, he came to see his travels as an education—in what helped a ranch thrive and what ultimately led to failure. He watched as some ranchers found ways to hold on to their land in the old, traditional ways while others saw the path to success as mechanizing, fertilizing, and spraying herbicides and pesticides. "When you work on all these ranches, especially the big corporate places, it's just soul killing," Hoy says. "What I saw were all these ranches just chewing up nature and spitting it out." He started thinking about running his own ranch and began to envision a different path, a simpler and more profitable one. He intended to clear out fences to allow open grazing, handle the cows on horseback, and avoid expensive chemicals. "I worked on several hundred ranches, and I bet there are only ten of them left," Hoy says. "They all had generational change or they went bankrupt."

Another decade passed before Hoy got the chance to put his ideas into practice. When he and a cousin pieced together a ranch in the Flint Hills, Josh and his new wife, Gwen, were determined to do things their own way. It has taken them most of the last quarter century to figure out what it means to work within the rhythms of nature. For many of those years, their experiments were greeted mostly with scorn from fellow ranchers. One fall day they got into a screaming match with a neighbor while buying gasoline for a controlled burn to clear their fields. The man threatened to call the sheriff: Ranchers were supposed to burn in the spring, period, and any other timetable was a crime. But as it turned out, the Hoys were merely ahead of the curve in adopting regenerative agriculture, a type of farming that can heal degraded soil and build biodiversity. Little by little they would manage to balance a mix of competing demands: how to produce beef in a profitable way, restore the grasslands for grazing, and recognize birds and other wildlife as an integral ingredient of a healthy ranch.

Remaking the Grasslands

Prairies historically covered nearly one third of North America, the natural grasslands spreading from the Rocky Mountains to east of the Mississippi River and from Saskatchewan south to Texas across half a billion acres. While creating an agricultural behemoth that feeds the world, the conversion of wild grasslands to mostly farms and ranches has claimed 60 percent of this native habitat that's home to 450 different types of grasses. In recent years, the United States has lost more prairie than the Brazilian Amazon has rainforest. Most Americans have no idea of the dimensions of the transformation still taking place, says Marshall Johnson, chief conservation officer for the National Audubon Society. "Something that eclipses the rainforest catastrophe is playing out in our backyard every year," he says. "There's a fundamental lack of appreciation for grasslands among most people."

If the prairie is easy to overlook, its place in the natural order is not. Vibrant grasslands create an ecosystem that benefits our daily lives. The deep, perennial root systems of these native plants sequester carbon and help recharge aquifers critical for irrigation and drinking water. They reduce soil erosion and filter nitrogen and phosphorus from agricultural runoff before it can overwhelm rivers and streams. Grasslands are vital for all sorts of pollinators, birds, and other wildlife.

Here again, birds are signaling the impacts of a deteriorating landscape. The causes range from the massive loss of habitat to widespread use of pesticides. The collapse of species in the nation's prairie surpasses all other habitats. The Three Billion Bird research showed grassland birds as a whole have lost 53 percent of their abundance in a half century. That translates to more than 700 million breeding individuals from thirty-one species. Put another way, nearly three quarters of birds by type in the Great Plains are in decline, often at startling rates. Eastern Meadowlarks, the voice of the grasslands with their plaintive song, are down by about 75 percent. The Greater Sage-Grouse, the symbol of the West famous for its theatrical mating dance, has dropped from some 16 million to somewhere under 500,000 birds.

The grasslands are a far different habitat for bird conservation than the forests, wetlands, and waterways of the East. Here birds live right on top of the proverbial breadbasket. As grassland conversion has reached epic levels, state and federal governments and nonprofits are looking for ways of preserving what remains. Conservation must mesh with the needs of private landowners on working farms and ranches that double as home to an array of sparrows, quail, grouse, and larks. A growing number of ranchers and farmers are gradually finding ways of maintaining open range, cultivating native plants, and rotating crops to create healthier environments.

We visited farms and ranches in California, Wyoming, Florida, and Kansas that are experimenting with variations on a theme: how to coexist with nature in general and birds in particular. In this chapter we zero in on three of them: the story of Josh and Gwen Hoy in southeastern Kansas; an experiment to save the Sage-Grouse as part of the most ambitious of the Department of Agriculture's conservation efforts; and the equivalent of an agricultural laboratory on a Florida cattle ranch. It's early days for many projects like these. But considering the sheer size of the landscape, this could be the most consequential place on the continent for protecting birds and helping restore a degraded environment.

Building a Sustainable Ranch

The first to greet us in the Hoys' driveway is a herd of enthusiastic family dogs, eight at last count. The couple's seventeen-year-old daughter, Josie, a home-schooled, sixth-generation rancher following in her parents' footsteps, quiets the pack. She leads us along a wraparound porch nearly bigger than the house, through a narrow mudroom, and into a roomy living space with soaring, twenty-foot ceilings. The family has recently moved into this new house, after faulty wiring had started a fire that destroyed their home three years ago. Mounted trophies of a bison and an antlered elk hang above our heads, and family photos show Josh's father and grandfather in their younger days wearing cowboy hats and leather chaps.

The Hoys' evenings center around cooking together, and a commercial stove takes up most of one wall. Just opposite is an island displaying a bowl of hummus from Josie's own recipe and a basket of flatbread she made from scratch. There's also the custard pie Gwen whipped up at dawn before a day on horseback herding cattle. The aroma of Josh's beef tips roasting with onions and peppers is seeping from the oven. We sit down at the well-worn wooden table at the center of their home to enjoy the meal and learn just how this bird-friendly ranch came to be.

It all started when Josh was speeding along the Flint Hills' Glanville Turnpike twenty-five years ago with his cousin Skip at the wheel. Josh spotted Gwen on horseback in a nearby pasture, working her afternoon job as a ranch hand, and he announced to Skip: "Oh, that's that new girl here who's, you know, cowboying. I'm going to marry her." Skip swerved off the road, laughing so hard the truck nearly ran into a ditch. Josh hadn't even yet met Gwen, who grew up an hour north in Burlingame on her parents' farm. Her family raised pigs and cows, and she loved to help manage them. But working as a cowhand wasn't a typical job for women then, Gwen says, so she followed her parents' guidance and got an education degree from Kansas State. After graduation she took a part-time job in the Flint Hills teaching ninth-grade English. "I thought, well, if I can't cowboy—you know that's what I really wanted to do—I'll just go someplace remote." She wasn't impressed with Josh at first. "I saw them that day, and I was like, 'Who are those fools?' They were hanging out of the truck window, waving." But that was indeed the start of a romance that led to marriage, just as Josh predicted. Gwen quit her part-time jobs to start working alongside her new husband.

They started their ranch, the Flying W, in the late 1990s after another of his cousins came into some family money. The three of them gradually put together 7,000 acres in a region of the grasslands called the tallgrass prairie. This ecosystem, named for plants tall enough to tickle the bellies of bison, once covered 170 million acres from Saskatchewan to Texas. But by 1950, two decades before Josh was born, most of it had been converted to wheat, corn, and soybeans. The last 4 percent of tallgrass prairie is located on ranches in the Flint Hills, where a unique geology of limestone and shale makes the soil too rocky to plow. "This is one of the most endangered landscapes in the world," Gwen says. "Both of us are pretty passionate about preserving it."

The choices loomed as soon as they got their ranch started. "I remember one of the first days we were taking possession of the place, the county extension agent rolled in," Josh says. "He gave me a whole list of the equipment and chemicals he said I needed to buy. I just kind of nodded and smiled and thanked him. I knew I wasn't going to do any of that." Instead the Hoys got busy in their fields, pulling out the invasive plants and coaxing back the native ones. While 80 percent of a tallgrass prairie's plants are grasses, some three hundred species of flowers or forbs, like bluestem, asters, and lupine, cover the rest. Since he was a teenager Josh had read books by Aldo Leopold, the naturalist and writer considered the father of modern conservation. What would it take to apply Leopold's philosophy of creating a cooperative relationship with nature?

One of the most important lessons Josh learned in his cowboy days was this: Maintaining and caring for a tallgrass ecosystem in a way that's good for birds and other wildlife is partly about choosing the right cows. The Flying W's herd is a different breed altogether from the 1,500- to 1,700-pound grain-fed cows on many ranches. The Hoys' grass-fed animals are leaner and smaller at about 900 pounds—"thrifty cows," as Josh puts it. Descendants of the Corriente cattle that can be traced back to the first cattle brought over from Spain in 1493, they're hardy, resistant to pests and heat, and efficient at both finding food and giving birth. "For years a lot of our neighbors have just been horrified at how ugly and small our cows are," Josh says. "They said we couldn't possibly make any money on them." But in fact, his thrifty cows make the ranch more profitable because they don't require expensive hormones, antibiotics, and grain. "It's all about the inputs," explains Josh. "The lower the inputs, the better chance you have of being profitable and sustainable."

The Hoys clearly love working as a family, and they talk over each other and finish each other's sentences. They explain how the cows roam at will over miles of open grasslands, grazing in herds like the bison did on the prairies of the past. "We can send them off and it can take two or three days before they see a fence," Josh says. "That really changes them. It's like kids getting to go out for recess." The cattle live in multigenerational herds where their offspring remain for longer than usual. "You'll see a lot of grandmothers with new mothers, and they'll help the mother raise the calf," Josie Hoy says. "It's just like parenting," adds Gwen. "They groom them and protect them. They learn what plants to eat, and where to go to graze." Grazing on a variety of nutrient-rich native grasses means the cows don't need antibiotics or other medicines. "They browse, taking bites from different plants," Gwen says. "It'll be different at different times of the year. They know what they need."

When a prairie is rooted in native plants, the fields become an ecosystem that the cows themselves sustain. "They're peeing, pooping, and slobbering, and that fertilizes everything," Josh says. "And every place the cow has stepped is a little indentation, and it's the perfect microbiome for the grass seeds to germinate." And the Hoys' cows are calm and contented, Josie explains, so the family can easily manage the herd on horseback with no additional employees or a fleet of vehicles. All of these cost savings mean the ranch gets a better return on investment while the native wildlife thrives.

The Flying W was finally on steady ground and Josh and Gwen were new parents in 2006 when tragedy struck. Josh's cousin and business partner died in a plane crash. And then it came time to pay "the death taxes," a crushing burden for most "land rich and cash poor" independent farmers, Josh says. The couple took out a $1 million loan to pay their portion of the tax, and the other heirs sold a sizable portion of the ranch to pay theirs. The staggering debt changed their way of life. In addition to herding cattle, they were suddenly herding visitors around their new guest ranch with Josh doing the cooking. They also took in 2,000 cows from other ranches for the summer grazing season, adding to the workload. The Hoys put easements on much of the property, which has emerged as an important part of conservation strategies for farmland as well as forests. Landowners such as the Hoys agree to refrain from development and maintain the land's zoning for agriculture in exchange for one-time payments that can reach millions of dollars depending on the location and amount of land. Through it all Gwen and Josh clung to their original conviction: Regenerative agriculture would be the only way to make the ranch profitable and allow them to safeguard their piece of this precious prairie.

If the Hoys once seemed like oddballs to neighboring ranchers, they're now viewed as models for building a business that coexists with nature. Three years ago, they won the Kansas Aldo Leopold Conservation Award, which honors pioneers in agricultural achievement on private land. Even more rewarding, they say, are the number of beginning ranchers who seek them out, often on Facebook. "We've been contacted by a lot of younger ranchers who are interested and very eager to learn," says Josh. "There's a groundswell out there."

The Hoys have teamed up with the National Audubon Society to produce "bird-friendly" beef in an experiment that enables consumers to support conservation with what they buy. The idea is that ranchers sell their beef at a premium of up to 70 percent in exchange for meeting a series of strict production and environmental standards that help birds. "We felt that one of the critical pieces that was missing from grassland conservation was market-based conservation," said Marshall Johnson, the Audubon chief conservation officer. "That is the ability to put conservation practices at the heart of the food chain."

A few years ago, the Hoys went into partnership with the Bobolink Foundation, led by Wendy and Hank Paulson, the former treasury secretary under George W. Bush. Based in Chicago and focused on conservation and biodiversity, the foundation purchased an adjacent 4,000-acre ranch that the Hoys use for open grazing while the foundation studies the impact on wildlife.

The native grassland on that ranch has now come back, and so has the wildlife. One of the most impressive signs has been the variety of birds. The land had long since lost the Henslow's Sparrow, a spiky-tailed bird with a subtle, olive green head, considered a barometer of grassland health. But in the foundation's recent commissioned survey, a handful of the sparrows turned up. "There was no hint of those sparrows out there before," says Justin Pepper, Bobolink's chief conservation officer. "Our assumption was that if we were able to bring the Henslow's back, we were restoring the habitat."

The Hoys recently reached a milestone they never thought they would. They paid off the last of their debt. Josh and Gwen even took a break from the Flint Hills. They spent two months helping on a ranch in Nebraska, which let Josie run the Flying W by herself. She has all the skills she needs to take over when her time comes, but she's not ready to decide if that's what she wants to do. As she turned eighteen, Josie decided to go to culinary school, and later would spend a few months ranching with friends in New Zealand. "Who knows what happens when you go on a walkabout," says Hoy, referring to the same sort of traveling he did at Josie's age. "But I suspect she'll be back. She loves us, and she loves ranching, too."

A Billion-Dollar Bird

Pineville, Wyoming

No bird in North America puts on a better show in the mating season than the male Greater Sage-Grouse, a flamboyant, turkey-like bird with a population scattered across the vast western landscape. Every spring, tourists who arrive at dawn at remote clearings, called leks, get to watch what amounts to an otherworldly bird orgy. The male's tail blossoms into a spiky fan and whisker-like strands rise up from the back of his neck. Downy white feathers cover his shoulders and chest, technically a ruff, but visually more like a decadent chinchilla coat. And now comes the unusual part. His chest puffs out, revealing two yellowish patches of bare skin that bulge like breasts from the plumage and make a popping sound. The bird starts thrusting suggestively, over and over, in a show we'd rate well beyond PG-13.

These birds may be looking for mates, but their outlandish show has ended up being a key to their survival.

The Sage-Grouse is a political conundrum. Its range spreads over 200 million acres in eleven states, much of it up against oil and gas fields, mines, and cattle ranches that the western economy depends on. The bird's population losses make it a candidate for the Endangered Species Act, but it's never been listed amid complaints that this would limit industry and agriculture. Every attempt over two decades to strike a compromise that protects both the bird and the local economy has fallen apart or landed in court. Each of the last three White House administrations swung back and forth on whether and how to protect this bird.

One solution started by the Department of Agriculture has slowly built momentum and sidesteps the politics. The idea is to incentivize cattle ranchers to protect the birds by focusing on where their populations are greatest. They called it the Sage Grouse Initiative and turned to University of Montana professor Dave Naugle a decade ago to work out the details. He realized that the leks pinpoint exactly where the most grouse could be found.

Every state wildlife agency tracks the bird's mating seasons, so they know just where the leks and populations are located. "We took each of those leks and did some fancy analysis." Naugle showed us a map that darkens to purple everywhere leks are plentiful and the populations are in need of the most protection. "It's an agreed-upon map, where everybody says, 'Yes, we're going to use these purple areas that are the occupied Sage-Grouse habitats.'"

The Sage Grouse Initiative faced plenty of skepticism. The pitch to the ranchers was they get financial support for making improvements to their land that supports the birds. But many weren't convinced they could trust a government agency to deliver on its promises. Environmentalists favor legal protection for the Sage-Grouse, not voluntary support. "I don't think it's a very good model," says Noah Greenwald, the endangered species specialist with the Center for Biological Diversity. Nevertheless, the Sage Grouse Initiative has grown into the largest of the Farm Bill's bird conservation projects by far. More than 1,900 ranchers have signed up for incentives that cover up to 75 percent of the cost of keeping sagebrush intact and pulling out invasive plants like red cedar, Ashe juniper, and mesquite that undermine the sagebrush. To date, the work has protected about nine million acres of sagebrush.

The project comes with a sizable bill: The Sage Grouse Initiative now spends about $50 million a year on ranch improvements. In addition, ranchers can get funding from development easements that keep prime land permanently in agriculture. All told, the initiative has invested more than a half billion dollars—with a nearly equal amount spent on the grouse by other state and federal agencies, nonprofits, and foundations. "Birders, ranchers, the government, energy companies—everybody's under the same tent," says Tim Griffiths, who leads the western branch of the U.S. Department of Agriculture's Working Lands for Wildlife program. "They don't get along on a lot of things, but they have a shared vision of having large, intact range land."

The population of the Sage-Grouse tends to rise and fall over time, so it's hard to pin down how well the initiative is working. The bird's numbers are still declining in many western states, but in the core population areas in Montana, Idaho, and particularly Wyoming, the populations are stable or rising slightly. That's critical since Wyoming is the state with the most birds as well as the largest oil and gas industry in the West.

Brian Jensen, a biologist who oversees the Sage Grouse Initiative in Wyoming, says the debate over the Sage-Grouse never stops. He hears criticism from those who think the Sage-Grouse needs more protection and those who think it already has too much. "I tend to think trying to strike some balance is the best thing," he says.

Jensen took us to meet Madeleine Murdock, one of the project's most stalwart supporters, who runs a 3,000-acre ranch overlooking southwestern Wyoming where 30 percent of North America's Sage-Grouse population is found. The rolling hills of sagebrush extend in every direction, midway between Salt Lake City and Yellowstone National Park, with snowcapped mountains ringing the valley almost entirely devoted to cattle grazing. Many of the ranchers have been here for generations, and they talk about the land and its wildlife with awe and pride.

"I think most ranches are happy to coexist with the native species, whatever they are," says Murdock, a stately woman now in her early eighties. "In many ways, they indicate the health of the area. So if you have deer and antelope and various birds, obviously they're finding enough to eat."

When Murdock signed up with the Sage Grouse Initiative, she got help to make improvements that protect the two dozen grouse on her land. For her, that means keeping the sagebrush itself healthy and making the ranch safe for the birds. She clipped reflectors on the fences so the grouse could see the wire and avoid collisions and built escape ramps inside the cattle's water wells so if birds fell in, they could waddle back out.

Like the Hoys, Murdock has taken out easements that cover about three quarters of her acreage as part of her estate plan to prevent development. The pressure on ranches isn't hard to see from Murdock's home, sited on the highest hill on her land. To her west and north, the rolling blue-green carpets of sagebrush extend for miles, dotted with a scattering of foraging antelope. But to the east, the town of Pinedale is marching toward her with the recent addition of a new subdivision that abuts her property line. As we drive the rutted roads along her ranch, Murdock says that after her husband died a few years ago she set things up so the ranch will continue under her son, an airplane pilot, and a ranch manager who's been part of the operation for years.

Just then, a Sage-Grouse scoots into the road a few yards ahead of us. It's such a common sight for Murdock that she hardly notices, but it is the first of these skittish birds we've seen. It's a hen, behaving as if a chick was nearby. The bird turns this way and that. With her small head and speckled brown-and-gray body, she's a low-key bird compared to the showy males at the leks. But she's still a striking sight against the green sagebrush.

Murdock's bird population varies with the years, she says, but the birds seem to be doing well. During the spring mating season, she loves stopping by the lek on her property to watch the mating performance. "It's a big thing for us to do, spend an hour watching them strut. How they developed this routine over presumably hundreds of thousands of years, we'll never know. But they're fascinating."

It's a glimpse of how nature has helped power her ranch through four generations of Murdocks. "If you lose diversity, you're affecting the whole balance," she says. "We want to see how we can help preserve this species, you know, their habitat. How we can coexist."

Making Room for Birds

Lake Placid, Florida

From an eight-foot-high vantage point looking across the Florida landscape, the last thing you'd expect to see are cattle grazing underneath palm trees. But that's exactly what more than 3,000 of them do here at Buck Island Ranch, near the center of the state outside of Lake Placid, and Hilary Swain couldn't be happier. She's executive director of the Archbold Biological Station, the premiere research center studying such species as the Grasshopper Sparrow and Florida Scrub-Jay. But the minute she steps out of her car at Buck Island, she becomes a rancher. And to see her eyes light up as the conversation turns to anything having to do with cattle, you know she's happy to be wearing a rancher's cowboy hat, too. "I love it because, here, I feel like I'm making a real difference."

Four years ago, Archbold, which had been leasing Buck Island, took a gamble and purchased the ranch to help demonstrate for Florida's ranchers how best to navigate the changing environment. Swain invites us to climb aboard a giant, open-air swamp buggy that looks like it's left over from a military invasion. And then off we go to see what a 10,500-acre science experiment looks like. Buck Island isn't only a research venture. It's a full-fledged cattle operation, and just like fellow ranchers, Swain must make sure Buck Island pays its way in this tough, low-margin business. "It's a commercial ranch. It's not some boutique type of thing," she says. "How can we have sustainable agriculture without diminishing natural resources? How can we produce food and make enough money running a ranch? What are the flows of nutrients? What are the flows of carbon? What is the biodiversity out there? How does all this work on working lands?"

Sarah Biesemier (left) and Juan Oteyza arrive on the Florida prairie with a carrying case full of Florida Grasshopper Sparrows.

Oteyza, a biologist with the state's Fish and Wildlife Research Institute, checks over a sparrow just before it's released into the wild.

Steve Shurter, chief executive at Florida's White Oak Conservation Center that's breeding grasshopper sparrows, with a white rhino raised at the 17,000-acre complex.

Tommy Michot, Peggy Shrum, and Steve Latta take a break in the Louisiana bottom-lands where they're searching for the Ivory-billed Woodpecker.

The Osprey lives almost entirely off its fishing prowess; it has surged in population along-side the Bald Eagle and Peregrine Falcon following the DDT ban fifty years ago. One of the symbols of recovery in the hemisphere, it's found in much of both North and South America.

Reed Bowman, a bird scientist at the Archbold Biological Station research center, has worked on rescue missions for the Florida Scrub-Jay and the Florida Grasshopper Sparrow. Here he's on the monthly census of the scrub-jays around Archbold.

Miyoko Chu, a biologist and director for communications at the Cornell Lab of Ornithology, oversaw the work to explain the Three Billion Bird report to the public. That's a Florida Scrub-Jay perched on her hand.

Mike Parr, president of the nonprofit American Bird Conservancy, pushed scientists to figure out how many birds have been lost over time.

John Fitzpatrick, the preeminent bird scientist who shaped the Cornell Lab of Ornithology over a quarter century, is now the lab's director emeritus.

Antonio Celis-Murillo directs the U.S. Geological Survey's Bird Banding Lab, which tracks the travels of the 82 million birds that have been banded in North America.

Sara Zimorski, a state biologist who leads the project to restore the Whooping Crane to Louisiana, wears a crane costume while moving among the birds.

Once too small for tracking devices, the Sanderling (weighing just two ounces) is now a subject of migratory studies using steadily smaller technologies to analyze the forces behind dramatic declines. Sanderlings, in the sandpiper family, can be seen skittering ahead of the waves on beaches around the world.

Sarah Sawyer (left) is the national wildlife ecologist with the U.S. Forest Service overseeing the California Spotted Owl project. Kevin Kelly leads the field work.

André Raine, a scientist in Hawai'i, hikes into the rainforest on the island of Kaua'i to check on a seabird colony.

15

The Palila, one of Hawai'i's honeycreepers facing the threat of extinction, is down to about 1,000 birds spread across the middle range of the Big Island's Mauna Kea mountain. Here it's feasting on the leaves of a Māmane tree.

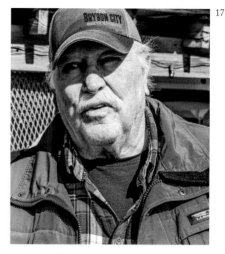

Ben Novak is the lead scientist with the genomic rescue nonprofit Revive & Restore and heads the team trying to bring the Passenger Pigeon back from extinction.

Richard Williams has been a guide for more than sixty years in the duck-hunting mecca of Currituck Sound in eastern North Carolina.

Douglas Linden, a wildlife biologist at Fort Benning, sets fires at the base of pine trees that are home to Red-cockaded Woodpeckers. The burning regime is critical to maintaining the woods around Fort Benning for the woodpeckers now making their way back from endangered status.

The Blue-chested Hummingbird is one of some 130 species of hummingbirds found among Ecuador's massive bird population.

The Dickcissel, whose song sounds like its name, can be seen throughout the continent's grasslands.

Josh and Gwen Hoy and their daughter, Josie (far left), operate the Flying W ranch in the Flint Hills of southeastern Kansas where they've established practices that support the bird populations.

left: Hilary Swain, director of the Archbold Biological Station in Lake Placid, Florida, looks over the Buck Island Ranch she oversees to study best practices in agriculture.

right: Pete Marra, who runs an environmental initiative at Georgetown University, helped write the Three Billion Bird paper and now leads the research project Road to Recovery.

above: The authors birding in a clearing in Central Florida.

right: In the Airstream, their home away from home during their research.

Buck Island is like a patient hooked to monitors that track its vital signs with projects set up on the grazing lands, wetlands, and canals. The ranch's runoff, which eventually winds its way to Lake Okeechobee at the center of the state and to the Everglades in South Florida, is analyzed for quality. Its species of birds, bees, and insects that help measure biodiversity are tracked. Researchers on the ranch are focused on the impacts of climate change, a topic obviously of deep concern in a low-lying state where water is everywhere.

Many of the state's ranches are owned by fifth- and sixth-generation Floridians who have vast experiential knowledge, Swain says, and they love their land and want to do right by the environment. Using the data from Buck Island's research, Swain and her staff can make recommendations that will help improve the land with sustainable agriculture, keep pesticides to a minimum, and boost economic productivity. "Our goal is to put our science into conservation action," she says.

Buck Island's success is convincing in itself. The ranch now ranks among the top twenty in Florida in terms of production, and its manager is a career rancher and a past president of the Florida Cattlemen's Association. And now ranchers from around the state are paying attention. Archbold and Buck Island share their findings in regular meetings with ranchers, conferences, and a steady flow of reports that go out to the Cattlemen's Association's 1,500 members. Swain sees ranches and farms as one of the central planks of the state's overall environmental strategy. "A conservation strategy that just protects public lands will never succeed, because it will not have the connectivity among those islands of public land," she says. That's where the massive acres of agriculture fit in. "Government agencies and conservation organizations are all very aware of this and are looking for ways of working very closely with ranchers."

The concept shouldn't be limited to agriculture. In the United States, slightly more than 60 percent of lands—forests, grasslands, wetlands—are privately owned, which means conservation on all private property is crucial. Amanda Rodewald, senior director at the Cornell Lab's Center for Avian Population Studies and who pays close attention to conservation trends, sees the idea catching on. "Traditionally, people thought conservation meant parks, protected areas, where you'd have the least impact on human activities," she tells us. "But clearly we need to expand our definition of what we consider conservation, or we're never going to make progress in a lot of places like the grasslands that are overwhelmingly privately owned. I do think our view of conservation has been changing over the last couple of decades, but it seems like it's really accelerating now."

Helping to prove the point, Buck Island shares one thing with all the ranches and farms we visited: Birds are everywhere, coexisting with the cattle, scattered throughout the fields, sharing the canals with turtles and alligators. The ranch attracts 177 different species, 90 of which breed here in these fields and waterways. We saw the same thing in Northern California's Davis Ranches, a rice farm on the Sacramento River where Dunlins, Kildeers, Black-necked Stilts, and egrets, ibis, herons, ducks, and geese filled the shallow ponds. On Michael Doane's ranch in central Kansas, he finds that bird populations are a measure of the health of the soil, plants, and air. "It's absolutely alive," says Doane, who helps run the family ranch and also works as global director of food and freshwater systems for The Nature Conservancy. "The birds are everywhere."

The loudest of all the farms and ranches we visited was Josh and Gwen Hoy's Flying W. We parked the Airstream in a lower field during our stay, filled with native grasses not far from a running stream. Every morning we awoke to a choir of sparrows, robins, finches, and especially Dickcissels, one of the prairie's most frequently heard summer sounds. The gray bird with brown-patterned wings and yellow highlights likes to perch on a stalk or wire and greet the morning with a simple song that sounds a lot like its name— *see, see, dick, dick, ciss, ciss, ciss.*

That's the same soundtrack that follows the Hoys around much of the day. "You hear a healthy prairie before you see it or smell it," says Josh Hoy. "You hear the bird songs, and you know you're riding in a pasture that's been grazed right. The healthier the prairie is, the louder it is."

Postscript: We're Not in Kansas Anymore

From Beverly:

Toward the beginning of our cross-country trip, I found an irresistible addition to our trailer's decor: Plastic placemats emblazoned with all fifty states. What a great way to trace our route day by day. The pace allowed plenty of time to relearn each state's shape, its capital, and a pertinent detail—peanuts for Mississippi, a cactus for New Mexico, potatoes for Idaho.

For me, Idaho was virgin territory, and I was excited to go. Over the next two days of late June we'd drive from Lake Tahoe to Bruneau Dunes State Park fifty miles outside Boise. It's home to the tallest single-structured sand dune in North America, rising 470 feet above the surrounding desert floor. The park looked peaceful and interesting on a state tourist website, so we decided to pause here for a few days of quiet, do some writing, and go birding.

I didn't know it yet, but the trip was about to take a turn.

If you want to experience the surface of Mars, Bruneau Dunes is just the spot. There was nobody in sight at the campground, only two other trailers and some Western Kingbirds perched in a leafless tree. Nearby, a yellow caution sign warned: "Sand temperatures can reach extreme levels." Not so long ago NASA scientists came here to study what conditions in the Mars dunescape might be like.

And this is where the Airstream finally let us down. With no trees and the sun beating on its metal roof, the air conditioner struggled to reach 87 degrees. So we decided to take a sunset drive to look for birds.

About halfway into the trip our cell phones started dinging with a weather bulletin warning of an approaching hailstorm.

For Airstream owners, hail is terrifying. Mere minutes of it can dent an Airstream's aluminum shell to the point of totaling the entire trailer. Should we turn around and go rescue it? Even if we did, there was no place in the desert to take cover. The skies above us looked fine, so we kept driving.

Roughly twenty minutes later the phones blasted again, this time with an alert for an approaching dust storm. Hail we knew; dust we didn't. I clicked on the accompanying link for Weather.com and found safety instructions. We were to evacuate the area immediately or else pull over to the side of the road, roll up the windows and hope for the best. Mars indeed.

In the end we found the Black-billed Magpie, whose wings and long tail turn an iridescent metallic blue-green in the right light, saw a gorgeous sunset, and returned to Bruneau Dunes unscathed. Neither the hail nor dust storm touched down where we happened to be. Even so, I felt like a bewildered Dorothy in *The Wizard of Oz*.

Thinking about our desert experience that night I realized my exposure to the American West—its deserts, prairies, and peaks—came mostly from movies. Hmmm. To what degree might my perspective on the rest of the country be this naive, and does it really matter? In the past I'd zipped in and out of unfamiliar cities on business trips and vacationed in scenic destinations. There was no effort to absorb the places in between.

The search for endangered birds, however, was a wholly different adventure. It required moving across those state-by-state plastic placemats a bit at a time. Birds kept taking us to places we never knew existed—to the West's sagebrush steppe, the tallgrass prairie of the middle states, hidden swamps in Louisiana. Along the way we started looking at the country from the perspective of birds.

From a bird's-eye view the story of the United States is about the land itself—millions upon millions of acres, one individual habitat after another. It's an ancient tale of birds evolving so closely alongside plants, bugs, and predators that even the shape of their beaks might depend upon a particular ecosystem. They simply cannot survive anywhere else.

Unfortunately for birds, a more recent chapter of this country's story includes the history of European settlers and how they acted upon the land. As Josh Hoy talked about learning to ride a horse in the Kansas tallgrass, you could hear how it grounds him, like maybe he'd have trouble living anywhere else. When he said only 4 percent of this prairie ecosystem still exists, I felt utterly sad.

This is a part of our country's history we heard in nearly every ecosystem we visited. The way it started out, each successive generation of pioneers simply did what pioneers do: They lived off the land, taking what they needed. But somewhere along the way "taking what you need" morphed into "grabbing all you want." Why didn't someone realize the resources were starting to run dry? Or maybe it's not possible to be a conservationist when nature pays the bills.

But what if it boils down to the fact that we as humans tend to love and perceive only the ecosystems that immediately surround us? I wonder what it would take for us to help each other care for all of them. Just like the birds, we simply cannot survive anywhere else.

Once thought doomed, the Red-cockaded Woodpecker has made a dramatic comeback with the help of the U.S. military, which has embraced its role as a caretaker for hundreds of troubled species.

10

CASE STUDIES IN GETTING IT DONE

Raleigh, North Carolina

The atmosphere in the Kerr Scott Building is part cocktail party, part carnival, which is fitting in this cavernous auditorium located at the North Carolina state fairgrounds. The master of ceremonies has his microphone turned way up and is encouraging everyone to check out the raffles, auctions, exhibits, and open bar. Old friends from around the state are shaking hands, slapping backs, and catching up on each other since last year's gathering. The early arrivals line up to buy a ticket to win a Beretta shotgun, which comes with a shot of Daviess County Bourbon. Others are going for the Camp Chef smoker grill or the fishing tackle. The occasion is Ducks Unlimited's annual statewide banquet in North Carolina's capital, drawing four hundred people, from lifelong hunters to first-timers recruited by the college chapters. "I think we do about 2,500 of these across the country every year," Will Johnson, the tall, easygoing development director for the Carolinas, shouts over the noise. "But we have north of 4,000 total events when you include the shoots, golf tournaments, and stand-alone raffles."

Ducks Unlimited banquets and tournaments are fund-raising forces—bringing in $50 million a year from dinners like this. But tonight's crowd that will contribute a record $187,000 isn't here solely to raise money. They're carrying on a tradition that blends networking, partying, and hunting that started in 1937 after the Dust Bowl nearly wiped out ducks and geese in North America. Since then, DU— as everyone calls it—has grown into the most successful grassroots conservation campaign on the continent, one that is passed along like an inheritance from one generation to the next. "You learn this from your father and his father before him," explains Bennett Whitehouse, an insurance broker in his early thirties from Raleigh. He's cradling a six-week-old Labrador puppy, one of the evening's raffle prizes he's showing around, as if he wouldn't mind taking the dog home himself. "Three generations of Whitehouses all go hunting together, which is really cool."

The success of duck hunters is the first of three case studies that look at broad, ambitious efforts to save birds. The second is the mission the U.S. military embarked on after confronting the problem of endangered species on its bases. The third is just getting started but is equally promising: Road to Recovery, an ad hoc volunteer group organized by the authors of the Three Billion Bird report. They insist a new approach to bird research is warranted to supplement a status quo that's falling short for the many species not yet federally protected but imperiled nonetheless.

Taken as a whole, these three examples show it is possible to accomplish what in today's world often feels beyond reach: provide the money, political clout, and scientific precision to save birds that are all too often about to vanish.

Case Study 1: Hunters Lead the Way

Conservationists trying to build public support for saving birds could do no better than follow the model Ducks Unlimited has created over the past eighty-six years. The nonprofit has 700,000 members in the United States, Canada, and Mexico, helping corral federal and private funds and establishing political clout unrivaled among environmental conservation groups. During his years as Ducks Unlimited's chief conservation officer, Paul Schmidt had only to lift his phone to apply political pressure on Congress. "I could call five to ten billionaires that I know of who were engaged in waterfowl, call them up on the phone and say, 'Could you call this senator or congressman and tell them you need more money for this or that?'" Schmidt says. "I can't do that for nongame birds, I'm sorry to say."

The hunting lobby has made the most of that influence, pushing through funding legislation and safeguarding the 10 percent tax on rifles, ammunition, archery equipment, and handguns. Since its passage in 1937, the Pittman-Robertson Wildlife Restoration Act's taxes have raised $14 billion for the conservation of waterfowl and other game animals. That money has mostly paid for land purchases and research. Each place setting at the DU banquet in Raleigh includes a map highlighting the land that Ducks Unlimited has protected for waterfowl. It adds up to 15 million acres in the lower forty-eight states along with just short of one billion acres under some form of protection in the boreal forest in the upper-reaching breeding grounds of Alaska and Canada. Waterfowl are the most watched bird group in the country, partly because they're everywhere and can be found in plain view almost any place there's freshwater—man-made ponds, lakes, rivers, and wetlands.

The duck strategies have created one of the few bright spots in the Three Billion Bird report. Game bird populations rose 56 percent—by far the largest increase—during the same time period many species have deteriorated. "I don't ever miss the opportunity to point out the game bird conservation that hunters have paid for has led to a remarkable recovery in the face of massive changes in society and development," says Adam Putnam, the former U.S. congressman and Florida agriculture secretary who's chief executive of Ducks Unlimited. "It's a proven model."

Ducks Unlimited draws admiration as well as jealousy from its fellow bird groups, including some who question why so much money and wildlife oversight goes to game birds in lieu of more troubled species. A better question is why haven't the other bird groups copied a version of what hunters have done—and is it too late to start? Several in the crowd at this banquet, concerned about the threats to all birds, say conservation groups are foolishly failing to put the Ducks Unlimited lessons to work. "The Audubon folks, The Nature Conservancy folks, particularly the birdwatching folks, they have a real problem with us killing ducks," says Brian Madison, an accountant from Winston-Salem, North Carolina, who's the DU volunteer state chairman for North Carolina for the year. "We're not the primary killer of the species. We're saving them. We need to bridge that gap and bring us all together so that we're on one mission."

For Ducks Unlimited, that mission started almost a century ago, three hours to the east of these fairgrounds, where North Carolina meets Virginia in the brackish waters of Currituck Sound. That's where Joseph Palmer Knapp, a magazine publisher and philanthropist from Brooklyn, first came up with the formula of modern duck conservation on a small island that's a haven for Mallards, Northern Pintails, Black Ducks, Ruddy Ducks, Buffleheads, Canvasbacks, and Blue-winged Teals.

The Spiritual Home of Ducks Unlimited

Knotts Island, North Carolina

Richard Williams started guiding hunters as a teenager six decades ago and has never stopped. He's spent his life watching the ebb and flow of waterfowl from his home on Knotts Island, this small, marshy spit of land that's like a hand reaching down from Virginia into the middle of Currituck Sound. Its geology created a rare ecosystem and some of the world's best duck hunting, nature's gift to a poor community with no roads and few job prospects. The original Currituck Inlet connecting the sound to the Atlantic Ocean closed up in a storm in 1828, filling the bay with mostly freshwater and its shallow bottom with tall reeds and cordgrass. "Ducks and geese followed in numbers that stagger the imagination," wrote longtime North Carolina environmental author Frank Tursi.

Forty-three hunting lodges sprung up along the sound, two of which were run by Richard Williams's family. By the early to mid-1900s hunters swarmed the region, among them some of the country's wealthiest industrialists drawn to hunting and whose largess would be instrumental in building the sport. They included the head of U.S. Steel, the founder of Eastman Kodak, members of the Dupont family and Joseph Knapp, who sunk the deepest roots of all into Knotts Island. Presidents Herbert Hoover and Dwight Eisenhower as well as Prime Minister Winston Churchill were among the world leaders who came here to hunt and fish. The wealth of waterfowl made the sound a perfect place not just for hunting but also for studying ducks. So when the game populations began to disappear following the drought of the 1930s and into the 1940s, Knapp decided to figure out what could be done.

Williams was just a boy in the last years that Joe Knapp had the three-story, thirty-seven-room mansion he used as a retreat to hunt and entertain friends from New York. In those days the island was reachable only by boat, and Knapp's impact was easy to feel even for a child. He created a school for islanders and gave every child shirts and either dungarees or a skirt as well as bags of candy and fruit at Christmas. Williams's great-uncle was Knapp's chauffeur and boat captain for years, and his great-aunt was Knapp's cook. But only after growing up did Williams realize the businessman's full influence.

Knapp ran the New York publishing house that included *Collier's* weekly and the Sunday newspaper supplement *This Week*. He chaired the Metropolitan Life Insurance Company, which his father helped start, and he invented a new generation of printing presses. But the work that mattered most on Knotts Island, and to duck and geese regions around the continent, was Knapp's role in the discovery of a methodical way to rebuild the duck population. He proved it was possible to study the birds, count them from airplanes, and understand what they needed in order to multiply. He secured habitat in their breeding grounds and along migration routes. He supported a duck research training center and created an organization called More Game Birds in America Foundation, which eventually became Ducks Unlimited. DU was formally launched in upstate New York. But as one writer put it, its spiritual home was Knapp's Knotts Island.

"My Lord. He was the backbone of the community," says Williams, now in his mid-seventies. "He made a big difference around here." We first met Williams one October morning when he was easing his sedan slowly along the sandy roads in Knotts Island's Mackay Island National Wildlife Refuge. Now preserved as part of the nation's wildlife refuge system, the marshes look much as they did in Knapp's day. Cattails sit atop stalks like hot dog buns swaying in the wind. Clumps of needlegrass and thick green reeds of giant cordgrass line the waterways. It's just a short walk from where Knapp's estate stood before it was leveled to create the bulk of the refuge. Williams patrols these back roads every day watching the ducks, songbirds, geese, and swans that stop over from one migration to the next.

Williams's years here reflect the growth and impact of ducks and DU on one of the nation's premier hunting grounds. He grew up between the two hunting lodges, each one started in the 1920s by a different set of grandparents. As a teenager during the academic year Williams spent several hours a day in a school bus—loaded onto a ferryboat—because that was the most direct way to reach the Joseph P. Knapp High School on the mainland. He'd already started guiding by then, occasionally producing a doctor's note to skip school in peak season. He watched as hunters from around the country came for the three- and four-day stays of early-morning hunts and country-style dinners. When the visitors pushed back from his grandmothers' feasts of roast beef, pork chops, chicken, and fish, they'd start in on a shot or two of whiskey and the storytelling that fueled camaraderie. "It was like Thanksgiving dinner every night," Williams says.

Though populations rebounded in the 1940s and 1950s, waterfowl remained susceptible to both extreme weather and the impacts of rapid development. The postwar boom gradually consumed about half of the wetlands across the country and by 1980, ducks, geese, as well as other wildlife were in jeopardy once again. In response, Ducks Unlimited helped put together the North American Waterfowl Management Plan, followed by congressional passage of the North American Wetlands Protection Act that provided funding for land preservation and conservation efforts. These expanded partnerships to protect ducks generated an estimated $6 billion in government spending and matching state and private dollars, and eventually added 30 million more acres of wetlands under protection. In Eastern North Carolina, that has translated into vast stretches of game lands, including nine National Wildlife Refuges scattered along the coast that have kept Currituck Sound and other inland waters healthy hunting destinations. The preserves are more important than ever now that marshes along the sound are slipping away at a rate of seventy acres per year due to development and sea level rise.

Hunters Funding Conservation: The Untold Story

Ducks Unlimited executive Adam Putnam said that what stands out about hunters is a willingness to pay their own way. "Hunters and anglers are frankly the primary funders of conservation in the United States," he says. "It's really an untold story. I think a lot of that is because, candidly, it's Ducks Unlimited, and it's considered a hunting organization, and I don't think people are going to give total credit for conservation achievement of that scale to a bunch of duck hunters."

Whether or not DU gets full credit, its peer groups recognize the hunters' impact and are openly envious of the revenues that taxes and hunting licenses provide for ducks. "Ninety percent of the money goes to 5 percent of the species," says Pete Marra, the Georgetown University scientist who helped put together the Three Billion Bird report. "And so one of our battles is how do we change that ratio and get 90 percent of the money to go to 90 percent of the species." In a later conversation, Marra said he's not suggesting that fees paid by hunters should be shifted to other birds. But he said the system built up over decades hasn't adjusted to the reality that while ducks prosper, 57 percent of other North American bird species are in decline.

Such an entrenched system isn't likely to pivot, especially since most conservationists historically have been hunters, going back to the presidency of Teddy Roosevelt, known for his love of hunting and creation of parks and preserves. The 567 National Wildlife Refuges built across the country continue to prioritize hunters and their quarry. There may be a butterfly trail in a refuge, but it's likely closed during hunting season. Many other bird species naturally benefit from refuge habitat, but rare is the staffer in a visitors' center who can tell you where in the refuge to find songbirds. (The exception is bluebirds—easy to spot in their wooden nest boxes.)

"That's because the history and the foundation of state wildlife agencies are rooted in game," Putnam says. "If your agency's charter is rooted in game, and your agency's funding is rooted in game, and the stakeholders who are most engaged in policymaking are game-focused, it's not any great surprise that the bulk of their priorities are related to wetlands and marine game species."

Will Johnson, the development officer with Ducks Unlimited, wonders why birders—who've never been as united or activist-oriented as hunters—don't recognize the simple concept of paying a price for your hobbies. "Who knows how many hundreds of thousands of people are harvesting pictures of warblers or Mallards or cranes," he says. "They're taking from the resources, but they're not giving anything back." He runs through all the potentially taxable items needed for watching birds: binoculars, scopes, tripods, travel expenses, and guides, not to mention feeders and an estimated $4 billion in birdseed sold in the United States each year. "None of that is ultimately going back to the source. It would be great if there was that same kind of commitment to the cause that hunters have."

Birding enthusiasts tried at one point to increase dedicated funding for nongame species. In the mid-1990s, thirty conservation groups came together to lobby for a 5 percent tax on a cross section of birding products, nicknamed the "binocular tax." The proposal came close to passage, but it was eventually blocked by lobbyists for the outdoor industry. Conservation groups then sought a way to share in the use of the Duck Stamp, the document behind hunting licenses that raises conservation funds for states. The stamp's illustrations are chosen from paintings by noted wildlife artists in an annual contest. Not only are the $25 hunting licenses a gift to conservation, the stamps typically become collectibles.

Schmidt, one of the top U.S. Fish and Wildlife leaders at the time, thought that alternating each year between illustrations of ducks and nongame birds would broaden the appeal and produce even more conservation dollars if the stamp appealed to a wider audience. "I thought, let's float this idea. I wanted to tear down this wall between the game and nongame folks," he says. "But it got shut down." Not only was the proposal rebuffed, the Fish and Wildlife Service staffers who run the contest went a step further. They passed a rule that future Duck Stamp images must include a gun or other hunting emblem. The directive didn't last long. Artists complained that putting shotguns or archery bows in such small renderings was awkward and unattractive. One entry included spent shotgun shells, which on the stamp looked like litter. But the message was clear: Hunters didn't want extraneous birds infringing on their turf.

The tension between hunters and birders seems to overlook how often their goals overlap. Ashley Dayer, a professor at Virginia Tech who works with nonprofits including Ducks Unlimited to widen support for conservation, says hunting and birding groups would clearly benefit from teaming up. "I think they're perfect partners," says Dayer. "They should be working together."

When a cross section of representatives from game and fishing organizations, bird groups, and business interests did come together in a push for better wildlife funding, it proved to be the most successful bid of its kind in years. A blue-ribbon panel, led by former Wyoming governor David Freudenthal and John Morris, who founded Bass Pro Shops, assembled a pointedly bipartisan group in 2014 to study how to correct the shortages and imbalances in wildlife funding. The panel's recommendations turned into congressional legislation called the Restoring America's Wildlife Act, which passed the House in June 2022, but remained stalled in the Senate in early 2023. The proposal would provide $1.3 billion annually for wildlife funding mostly through state agencies. It's the first attempt in years to deliver new money for birds and other wildlife especially aimed at troubled species of all kinds. Supporters still hope the legislation may advance and demonstrate the power of combining forces for birds. "I think this is a noble effort," Freudenthal tells us. "I do think this begins to reflect the notion that the traditional divide between sportsmen and the others is going to have to be bridged."

Case Study 2: The Military and Its Woodpecker

Columbus, Georgia

Georgia's Fort Benning, a hundred miles southwest of Atlanta, is where the U.S. Army trains more than half of the nation's new recruits. In any given week, the base swarms with 100,000 troops and civilians, and everything here is aimed at preparing the soldier for combat. Multiple rifle ranges and drop towers for paratroopers are located just off the main entrance of the base. A mile away there's a full-scale replica of an Afghan village where soldiers learn to enter, search, and shoot to kill. Traffic anywhere on Fort Benning's 182,000 acres can slow to a crawl behind marching platoons carrying full backpacks and rifles at their sides. In addition to basic training, troops are learning to drive the space-age version of an armored tank, becoming elite Army Rangers or attending Officer Candidate School.

Regardless of their course of study, mission, or rank, everyone stationed at Fort Benning confronts this salient fact: Soldiers may be the priority, but they'd better not pick a fight with the Red-cockaded Woodpecker.

The woodpecker, a black-and-white bird named for the nearly invisible streak of red toward the back of its head, is the symbol of one of the Department of Defense's least-known accomplishments: bringing endangered birds and other wildlife back from the edge of extinction. The U.S. military has been forced to play nursemaid to more endangered species, and at higher densities, than any other agency. That's because as the United States developed an ever-growing percentage of its forests, grasslands, coasts, and wetlands, the nation's five hundred military bases have remained islands of prime habitat harboring all sorts of rare plants and animals.

The Department of Defense has gradually built a network of scientists to research precisely what each species needs to survive, and every base employs environmental teams to implement the findings. With millions to spend and little bureaucracy to contend with, the DoD is a model for how to get the job done—in true military fashion. At its Seal Beach weapons station near Los Angeles, the navy has raised the elevation of marshes for the Light-footed Ridgway's Rail where rising sea levels are flooding the bird's breeding grounds. At Utah's Dugway Proving Ground near Salt Lake City where the army tests biological weapons, military pilots steer drones along the impassable cliffs to watch over its Golden Eagle nests. The army is installing bioacoustic monitors at its training base on the Big Island of Hawai'i that trigger an alert so planes can be diverted if Nene geese fly too close to the runways. In an ongoing project to recover the Golden-cheeked Warbler at Fort Hood near Killeen, Texas, biologist John Macey puts geolocating backpacks on the warblers to figure out where they get into trouble during long-distance migrations.

The bird that forced the military to embrace its natural resources mission in the first place is the Red-cockaded Woodpecker, a species that's spread throughout fifteen southeastern military bases. The woodpeckers nest in longleaf pines nearly a century old, and at Fort Benning, many of these trees are adjacent to the bombing range. Explosions rumble across the base with blasts so powerful they can shake the woodpecker right off his tree. Barely fazed, the spunky bird pops right back up on the tree just minutes later, 4.4 to be exact, according to one study. "A lot of people think, 'Oh, it's endangered,' and they associate that with fragile, and this bird is not," says James Parker, Fort Benning's chief of natural resources management. "It's a tough bird." Parker, now in his mid-forties, has worked on the woodpecker since graduate school, and he was surprised to find an unlikely scenario developing at Fort Benning. "It turns out what the Red-cockaded Woodpecker needs works right in line with the military mission."

This was not always the case. It took decades after the Endangered Species Act became law for base commanders to accept what they were up against. Fort Benning was bulldozing through a woodpecker nesting site to build a barracks in 1976 when a scientist came to consult with the colonel in charge. "I ain't never seen a red-cockadoodled woodpecker, and I never want to see one," the colonel told his visitor. "You do what it takes to move them." By January 1992 the conflict at Fort Benning had moved to the courts. A federal grand jury indicted three civilian employees for conspiring to "take" (harm, harass, move, kill, etc.) Red-cockaded Woodpeckers and then lying about it to criminal investigators. "They had a hiccup," is how Parker puts it.

At about the same time at Fort Bragg in North Carolina, the hiccup sounded more like a belch. Training officers were caught treating the precious nest trees like concrete poles. Soldiers drove tanks around the pines and installed big guns directly underneath them. They tied cables around trunks that were clearly painted with orange circles indicating active nests overhead. When the U.S. Fish and Wildlife Service found out, Fort Bragg's team refused to discuss the recovery guidelines it had previously agreed to and issued a belligerent report demanding to train soldiers any way it saw fit. Then they staged a massive training exercise involving seventeen artillery battalions that heavily damaged the woods where the woodpeckers lived.

For the Wildlife Service officers, that was the last straw. Using the powers of the Endangered Species Act, they immediately shut down basic training at the largest military base in the country. Shocked military leadership did a permanent about-face, making changes that eventually led to a sweeping environmental policy for protecting biodiversity on all military installations. "Across the army as a whole, it was kind of the starting point for really gearing up for active management practices," says Tim Marsden, an army biologist at Fort Benning. The Department of Defense now has two separate research departments under its natural resources branch with an annual budget of about $25 million for researching conservation issues, environmental innovations, and climate-caused disruptions at its facilities. But the Defense Department doesn't sugarcoat the fact that the underlying reason for all its environmental work is to ensure a high state of military readiness and support national security. And a key goal is to avoid limits on training and weapons testing that can be imposed under federal law to protected listed birds on the bases. "The Department of Defense is not a conservation organization," says Ryan Orndorff, director of natural resources for the DoD. "Everything we do, all our programs, exist to enable our mission."

A Very Picky Bird

By mid-morning flames are licking at the base of one of the Red-cockaded Woodpecker's nest trees. It's a windless April day at Fort Benning, perfect for setting the woods on fire. Staffers are zipping around on ATVs, some squirting a slow-burning mix of diesel fuel and gasoline on pine needles and bunchgrass to coax along the flames. Others drive behind carrying water tanks, just in case. Meanwhile, as usual, the woodpeckers have flown out to look for food. By the time they return in late afternoon, all traces of flames will be gone and their territories will be trimmed up just the way they like it.

Every three years, foresters usher these fires through each woodpecker area to rejuvenate them. The controlled burns mimic fires that used to be caused more frequently by lightning strikes. It can be dangerous work. One of Parker's staff had to be airlifted to a burn center by helicopter after her clothes caught fire. Another broke his back when his ATV hit a cable and he flew backward. Parker himself sustained third-degree burns while using a torch to light fires some years ago. "It flashed and burnt my whole face," says Parker. Despite a crusty face and missing eyebrows, he was back at work the next day.

Fires are unpopular in residential areas around the base because of smoke and the threat of wildfires. While many government agencies are hesitant to upset the public, Fort Benning has no choice but to burn. Here's why: The Red-cockaded Woodpecker's favorite tree is a living longleaf pine. In an all-too-familiar story, the vast longleaf forests that once covered 92 million acres from southwest Virginia to eastern Texas are nearly gone—except on southeastern military bases. For some sticky reasons we'll get to, the tree is of no use to the woodpecker until it's at least eighty to a hundred years old. And the longleaf pine can't live that long without a landscape cleansed by fire.

Longleaf pines are unusual trees for a bird about the size of a cardinal to manage. If the trees are attacked by bugs or otherwise punctured, they ooze copious amounts of sap to seal the wound. For the Red-cockaded, this sticky sap has pros and cons. The bird's main predator—the rat snake—can't climb a sap-covered trunk. Thus, after carving out a nest cavity, the crafty bird pecks some holes around the opening, causing syrupy sap to form the protective barrier. As nature would have it, however, that same sap can fill up a nest. Consequently, the woodpecker works a little at a time, then goes away until the sap dries. The bird continues working through the layers of "sapwood" until reaching the heart of the tree. When a longleaf pine gets old enough, its heart starts to die and no longer produces sap. And so, this is where the woodpecker locates the nest cavity.

All other woodpeckers in North America build nests in dead trees in around two weeks, but it takes Red-cockaded Woodpeckers a decade or more. To make matters worse, every bird in the family group, called a cluster, must either build a cavity or inherit one. It looked like the woodpecker was doomed once most of the pines were gone.

The solution came from a North Carolina State University professor who spent his early career studying what makes the bird tick. He's part of a group of academics dubbed the "woodpecker nation" because so many have spent their entire careers trying to solve the mysteries of just one bird. Jeff Walters and his graduate students realized they needed a way to speed up the nest building if the woodpeckers were going to survive. In the late 1980s, they decided to experiment with drilling cavities at just the angle and height the birds like. When they tested the idea, the birds instantly set up housekeeping. Walters will always remember the day the student who led the project called him from the field with the news. "It works," she yelled over a pay phone. "It actually works."

Then, just as Walters and his students were completing their experiments, Hurricane Hugo swept through the South in September of 1989. The storm killed dozens of people and caused some $10 billion in damage to homes, buildings, and much of what remained of the longleaf pines where the woodpeckers congregated in South Carolina. "They lost 90 percent of the cavity trees," says Walters, now graduate director of the Department of Biological Sciences at Virginia Tech. "They got snapped right off, just knocked over." By this time Walters's team had figured out how to build artificial cavities that could be installed safely in younger longleaf trees and also in mature loblolly pines. They rushed to help the Forest Service put in hundreds of the inserts in hard-hit areas. "And it did keep the birds, I would say on 70 percent of the territories, despite having only 10 percent of the trees left," he says. "The populations grew back to where they were before and in not a very long time period."

The hurricane helped to jump-start the overall rescue plan, and the military embraced the inserts, installing these tiny clusters of government housing on bases wherever the birds needed them. For decades, Fort Benning has planted longleaf pines to provide for the birds of the future. Today, 412 woodpecker clusters nest in 1,276 trees, and James Parker's staff of twenty-four specialists watch over all of them. If a training maneuver needs to happen within two hundred feet of a nest tree, commanders must first ask the specialists' permission.

They keep an especially close eye on woodpeckers living near the bombing ranges, and they watch every spring when a new generation of woodpeckers arrive. They'll climb to the top of these trees when chicks are first hatched so they can reach in, fool them into thinking it's a parent, and slip on numbered bands, the equivalent of dog tags. When construction on the base may endanger the birds, they'll move the cluster of woodpeckers to a new location out of harm's way. They compare notes with other bases managing woodpeckers to learn what works best.

The Red-cockaded population at Fort Benning had now stabilized to the point that the birds carved out many of their own cavities and were thriving, says Parker, a natural storyteller in true southern fashion. So it seemed the base had put its initial disregard for woodpeckers to rest. But then came the Friday afternoon of a Memorial Day weekend a few years ago when firing tanks caused a wildfire on the training range. Parker's staff rushed to the scene, ready to extinguish the fire before it reached the nest trees on either side. But the soldiers were preparing for deployment, and the officer in command refused to stop training so the foresters could go in for the rescue. That was a choice they would soon regret.

"They Died of Smoke Inhalation"

In the earliest days of its life a Red-cockaded Woodpecker chick's job is to stay put and wait for food. Once the chicks are four days old, adults begin leaving the nest to bring back a meal of insect larvae, wood roaches, or centipedes. Meanwhile the featherless baby birds haven't yet opened their eyes, and they huddle together facing one another, each with its neck propped up on its siblings. They stay this way, making a soft rhythmic sound called "contentment peeping" until their parents return and they launch into begging mode.

That's likely what the babies were doing in their nests on that Friday afternoon as the flames crawled toward them. By the time the training was over five hours later, it was too late. When biologists got to the scene they retrieved the remains of two chicks and put them in a freezer for safekeeping. Parker was in charge of the required protocol. "Anytime I have a dead Red-cockaded Woodpecker, I have twenty-four hours to let the commander and the U.S. Wildlife Service know," he says. And so began another hiccup at Fort Benning.

The following Tuesday morning, Parker was in his office when a black SUV with tinted windows pulled up. Out stepped a federal investigator who'd driven down from the Atlanta U.S. Fish and Wildlife headquarters unannounced to sort out the damage. "So he starts demanding, 'I need this. I need to talk to these people.' So I said, 'Hold on, dude. You're on Fort Benning. There's a process to this.'" That started weeks of interviews and collecting evidence. "The investigator actually had an autopsy done on those chicks," Parker says. "They died of smoke inhalation."

While the Wildlife Service was starting to investigate, the army launched its own inquiry. "The colonel was furious," Parker says. "He's furious that we were denied access in an emergency situation, and he wanted to know why." So both teams of investigators sat side by side, talking to everyone involved, recording the conversations and putting it all together in a report that found the training command hadn't followed its own policies by refusing the emergency rescue. The garrison commander called it a serious lapse in judgment. "They'd made a very bad decision," Parker says. "He called in the several people involved and said, 'You can either retire, or quit, or I will fire you.'"

Parker now has the commander's cell phone number on speed dial in the event of another emergency. "I can call him right now and I bet you he will answer no matter what he is doing," Parker says. "But it has never gotten to that point again because everybody knows what happens." If the natural resources staff calls for an emergency training halt now, Parker says it is immediately granted. "We're here to do the right thing," he says. "Sure, they had a hiccup years ago. We try not to have hiccups now."

The Red-cockaded Woodpecker was the impetus for a grand and rare celebration Parker helped manage at Fort Benning in September 2020, and top administration officials were out in force. Before a row of American flags, the secretaries of the interior and agriculture joined the director of the Fish and Wildlife Service and base garrison commander Colonel Matthew Scalia to celebrate a true victory. The Red-cockaded

Woodpecker would be moving one step down the ladder on the Endangered Species List, from endangered to threatened status, years ahead of schedule. "This is a tremendous success story," said Interior Secretary David Bernhardt. "This is very, very significant."

Like almost everything about this bird, the change has come with plenty of complications. Two years of haggling over the specific instructions of its down listing followed the Fort Benning announcement, and hundreds of comments—many of them taking issue with the decision—flowed in from the "nation" of researchers who study the bird. Others worry that without the high-level endangered status, the overall federal funding needed to plant trees, maintain cavities, and do controlled burns won't be guaranteed. It's a significant investment. In fiscal year 2017 alone for example, the Fish and Wildlife Service reported more than $12.5 million in federal funds were spent on Red-cockaded recovery—part of the $79 million the four branches of the military spent that year on all its listed species.

But Fort Benning will continue the same routines for the bird's care, and the conservation budget will be secure as well, says Parker, who's been working on the project for twenty-two years. There's a point of pride to protect: Fort Benning's woodpecker population has far exceeded the numbers outlined in the bird's federal recovery plan and is twenty-five years ahead of schedule. The population is now growing by roughly twenty new woodpecker trees each year. Caring for this bird has become a way of life at Fort Benning.

Orndorff, the Department of Defense's head of natural resources, says the woodpecker has taught the bases a valuable lesson in conservation standards that don't just help the birds but improve the land in everything from training maneuverability to the quality of the environment for troops. "That's what the Red-cockaded Woodpecker has really done. As painful and as long-term a process as it's been, it's really helped us to define how we merge the benefits to the species with the benefits to our mission," he says.

Case Study 3: Rethinking Conservation

Washington, D.C.

Two months after the publication of the Three Billion Bird report in *Science*, some of the authors met for dinner in January 2020 at the Sovereign, a Belgian pub in Georgetown. Pete Marra, Ken Rosenberg, and Tom Will were still savoring the encouraging response they'd gotten to their research. But there was an uncomfortable flip side to their conversation in the wake of the revelations. Each had spent their careers trying to save birds, so the magnitude of losses they'd uncovered haunted them. Marra gave voice to what everyone was thinking: "You don't just go back to your day job after publishing something like that," he said. "We've got to rethink what we're doing."

The scientists weren't entirely sure what that meant, but the details began to take shape during a brainstorming session with a dozen peers in Marra's office the following day. Almost everybody came from different institutions—Marra at Georgetown University, Rosenberg at Cornell, and Will with the U.S. Fish and Wildlife Service in Minneapolis—and they had no authority or budget of their own to take immediate steps. At some point during the discussion, Will walked to the whiteboard and started taking notes. Soft-spoken, thoughtful, and a good facilitator, Will began to capture pieces of what they might do. "There was this explosion of ideas," Will says. "'We could do this. We could do that. What if we tried this?' We were trying to get some focus to it."

What they came up with was a new, loose-knit organization called Road to Recovery, or R2R for short. In the coming months, they would go on to identify fifty or so bird species that need in-depth research as part of a renewed emphasis on science and the fresh technologies of bird study. The mission is to get out in front of the potential avalanche of troubled species they see coming, and target birds approaching what they call "the tipping point" while they're still strong enough to recover.

Road to Recovery is out to do more than elevate scientific research. It has also served up a challenge to a field its organizers think needs reform. R2R is the equivalent of a start-up looking to disrupt what they see as a staid and passive discipline failing to respond to a crisis. "What's clear to me is something is not working," Marra says. "People may not want to hear that, because they've devoted their lives to conservation, and I'm sorry, but so have I. And what we're doing is not effective." The challenge resonated with many others in the bird sciences: Within months, hundreds of scientists across the United States, Canada, and Mexico volunteered to sign up for R2R workshops exploring how to make better use of technologies and how to get the field working together.

Road to Recovery also enlisted Ashley Dayer, the Virginia Tech professor who's been pushing for new approaches in conservation that do a better job of recognizing the public's role in the cause. "We're constantly hitting our heads against the wall when it comes to working with people on these issues," she says. Her role is to help scientists learn to break out of the myopic focus on their research and recognize the part average people play in conservation. "We're not looking for birds to change their behavior," Dayer says. "We're looking for people to change their behavior."

As their initiative gained momentum, the growing volunteer team of scientists talked about the degree of difficulty they face. It's not possible, they concluded, to restore the vast abundance that once characterized North America's birdscape; too much of the required habitat is gone. "The dirty little secret is that we're not going to bring them back," Marra tells us one night over dinner in Washington. "We're hoping we can stabilize them. I mean, you're not going to reforest Kansas and Oklahoma. You're not going to reforest parts of the eastern corridor." Instead, he envisions a slow and steady exploration of what is holding back specific species in order to first halt the declines, and then perhaps make targeted gains in populations. In some instances, the revival of bird species may look more like zoos in the wild, where birds are cordoned off as with the Hawaiian seabirds. Scientists will have to care for them indefinitely in a kind of rehabilitation zone similar to the section of the Florida prairie where researchers watch over the Grasshopper Sparrows like fragile patients.

Road to Recovery has a delicate balancing act to manage. Its leaders are part of the status quo, but they're critical of their own traditions. Marra has emerged as the chief advocate, and he creates a sense of rebellion. He's a respected scientist who's outspoken, blunt, and doesn't contain his frustration.

He questions why existing organizations aren't leading a push for fresh answers in the wake of the *Science* study. "National Audubon could have done that, but they didn't," he says. "Fish and Wildlife could have done that, but they didn't." He's especially tough on what he views as the federal agency's lackluster attitude. "There's been no response. I mean, what do you see Fish and Wildlife doing?" he asks.

It's a fair critique: Not only has the agency done very little in the wake of the loss of billions of birds, it has downplayed the significance of the findings. "I don't get excited about the number because that happened over fifty years," says Jerome Ford, the assistant director for migratory birds, one of the longtime leaders who speaks for the agency. "Our role was to calm everybody down." Fish and Wildlife has placed its emphasis on trying to stop collisions with high-rise glass buildings. But it has taken no leadership role, a stance that even many of its own former top officials find frustrating. Jamie Rappaport Clark, once a Fish and Wildlife director and now running the nonprofit Defenders of Wildlife, says the service is underfunded and faces political pressure from all sides. "They've been really beaten down," she says. Don Barry, the former chief counsel for the agency, adds: "The unfortunate thing is that the Fish and Wildlife Service has increasingly ceased enforcing the teeth of the Endangered Species Act."

At a time when the number of species in need is rising, federal funding is falling steadily further behind. The most detailed study of Endangered Species recovery spending found the Fish and Wildlife Service is unable to address the great bulk of the birds, other wildlife, plants, and insects reaching endangered territory. The agency's total recovery funds in 2016, the year of the study, were $82 million, which barely covered the administrative costs. The report, conducted by the environmental watchdog nonprofit Center for Biological Diversity and titled "Shortchanged," concluded it would actually take $2.3 billion annually over ten years to restore the 1,500 species of all kinds currently on its list—not to mention the hundreds more backed up waiting for needs assessments. "The Endangered Species Act is not failing," says Clark. "It's starving."

Like many in the field, the Cornell Lab's John Fitzpatrick says supporters of conservation as well as the service itself fail to make a strong enough case for adequate funding. "Why aren't we making better arguments, much more strenuously and vociferously, to the people that matter, instead of noodling around the edges of the crumbs on the plate and pretending that that's all we get?" Fitzpatrick says. "That, to me, is a profound issue in conservation."

The Road to Recovery leadership is also critical of the haphazard and sometimes contradictory way bird research is funded and directed. The federal government itself has gradually stepped out of the role of researching birds, cutting back staff and funding under the U.S. Geological Survey. "They're a shadow of their former selves," Dan Ashe, the former director of the Fish and Wildlife Service, says of what remains of the federal research staff.

That has left most of the scientific research scattered among nonprofits like the Cornell Lab and National Audubon and also university professors who seek out private and federal grants. The National Science Foundation, the primary federal funder, has not been supportive of the kind of practical conservation research that R2R envisions.

The academic research model creates a vast amount of scientific data, but many researchers say the academic journals that shape the science through what they publish aren't focused enough on applicable solutions.

"It's driven by the need to publish instead of the need to answer conservation questions," says the American Bird Conservancy's Mike Parr. Many scientists we spoke with had the same concern. When we ask University of Montana wildlife biologist Dave Naugle how much academic research leads to practical, on-the-ground conservation, he shakes his head. He holds his finger and thumb a fraction of an inch apart. "It's just not happening," he says.

Road to Recovery has succeeded in its initial goal: It's gotten the attention of the scientific community and raised the first $1 million of the $50 million it plans to invest in research over time. It has increased the number of birds it plans to study to ninety that are in decline but plentiful enough to recover. It's launched research initiatives for the first four species—the Evening Grosbeak, Golden-winged Warbler, Yellow-billed Cuckoo, and a shorebird called the Lesser Yellowlegs—to illustrate what R2R can accomplish as part of its funding appeal.

These four reflect the gamut of the challenges facing the entire collection of birds the group hopes to address. The Golden-winged Warbler, for instance, is an eastern forest bird that, similar to the Cerulean, has lost most of its population as the woods have changed and undergone development. The Lesser Yellowlegs—named for its stalky yellow legs—is the most hunted shorebird in the Americas for its meat and faces evaporating coastal habitat. The Evening Grosbeak, with pronounced yellow, white, and black markings and a sweet, piercing call, is a mystery of a bird. It's been left out of in-depth study even as it has lost more than 90 percent of its population. In this sense, it represents the wide gaps in knowledge that work against the rescues of many species. The cuckoo is another long-distance migrator running into trouble all along its route that stretches from the bottom of eastern Canada to the top of Argentina.

Road to Recovery hired Paul Schmidt, a calm, low-key veteran of both the Fish and Wildlife Service and Ducks Unlimited, as its director to build a staff and create partnerships. He says these first research projects are pushing hard to report back within one year, with at least initial assessments of what the birds need. Each species has a working group of researchers who have sometimes years of experience with these birds, and each has been awarded down payments of $100,000 to get started, mostly provided by the Knobloch Family Foundation, which is devoted to funding conservation science.

The working groups have all launched tracking projects to look for obstacles along migration routes and have probed the species' biology, mating, and nesting routines. Schmidt says the response among these scientists, as well as the wider group of specialists who attend R2R workshops and contribute ideas, has been encouraging. "There's a tremendous thirst for more workshops, more engagement, more training, and more networking with people addressing these challenges," he says.

R2R is still a small, privately funded organization that is cobbling together an ambitious undertaking from the ground up. The true test won't come until the first phases of research are complete and it has identified the most promising conservation tactics for each bird and looked for overarching trends. That's the point at which this start-up effort will need to enlist existing organizations and government agencies, philanthropies, and conservation groups in other countries to join the rescue campaigns. That's the day Marra is looking forward to.

"We're hoping this will build enormous buzz, and let us tell people about these species in ways that create a real swelling of interest. And maybe there will be other organizations that join us that are going to take this bull by the horns and move things forward." At least up to this point, Marra thinks the timing is right for real change. "It's a moment to turn things around. It's a real moment," he says. "This is a moment we can't let pass."

The Ruby-throated Hummingbird delivers one of the core services that birds provide in the balance of nature. Hummingbirds alone help pollinate more than 8,000 species of plants and flowers in North and South America.

MAKING THE CASE FOR BIRDS

Birds have always struck a special chord with people. They're among our most enduring cultural symbols, inspiring movies, novels, paintings, even the names of cars—from the Ford Falcon to the Plymouth Roadrunner. We name our sports teams for birds, including 1,300 college, university, and high school teams that call themselves Eagles. Mahalia Jackson singing the cherished refrain of the timeless gospel hymn—"For His eye is on the sparrow, and I know He watches over me"—has stood the test of time, winning the Grammy Hall of Fame Award in 2010, fifty years after she first performed it. A Dutch painting, *The Goldfinch*, became the inspiration for Donna Tartt's Pulitzer Prize–winning novel and its film adaptation. This 1654 painting by Carel Fabritius depicts a creature that's supposed to fly free but is chained by its leg, turned into a decorative pet. The consistency of bird songs inspired Emily Dickinson's 1861 poem "'Hope' Is the Thing with Feathers." Its brief three stanzas forever entwine the notions of birds and hope.

People have always envied a bird's freedom and ability to fly. Watching birds fascinated and enlightened the Wright brothers as they built the first airplanes. Even today aerospace engineers study the mechanics of how birds fly to design safer aircraft and shorten landing distances. Studies of gulls and hummingbirds guide the creation of more efficient drones and robots.

Birds make a vast range of contributions, some obvious, some still only partially understood. Advances in research on birds have helped document their place as nature's workhorses, handling such duties as pollination, seed dispersal, fertilization, and soil formation. These roles are vital for a healthy environment, and according to one detailed assessment, the global value of wildlife biodiversity in everything from natural services to human outdoor recreation reaches tens of trillions of dollars. A tougher calculation is what happens if these contributions gradually diminish or the bird populations disappear altogether. Scientists cannot say for sure what the results might be in part because of how deeply interrelated nature's fabric is. They don't know what happens as strands of this web are pulled out. But they fear cascading impacts of the erosion of these natural processes that in effect help to power and lubricate the earth.

What scientists can pin down is how valuable individual elements are. More than 8,000 species of plants and flowers in North and South America, for instance, rely on hummingbirds for pollination. One Dutch study found that where birds are around to consume insects feeding on crops of apples, productivity goes up 66 percent. Because of their mobility, birds are exceptional vehicles for handling the long list of chores that otherwise wouldn't get done. Their poop provides the best way of circulating nutrients throughout the oceans and islands. About 20 percent of bird species also spread seeds, which move intact through their systems, and are deposited far and wide as birds fly about.

On Hawai'i's Kaua'i, researchers are working to keep the last few hundred fruit-eating Puaiohi thrushes safe, in part for fear of what their disappearance would mean for the dynamics of the rainforest. These brown, nondescript thrushes, hardly ever seen, spread seeds they eat around the island in their daily travels. "We have no forest without the Puaiohi," said Cali Crampton, the head of the Kaua'i Forest Bird Recovery Project. "Without forests, we have no flood control. We have no drinking water." Hawai'i has already experienced life without the beauty and inspiration birds provide. In the mid-1900s when avian malaria wiped out most native lowland birds, the islanders couldn't stand the eerie silence. A club of Hawaiians went on a shopping spree and imported species from around the world to replace their missing birds. These stand-ins remain the bulk of the birds that tourists see across these islands.

Vultures, condors, and many raptors handle a nasty job critical to the environment: They clean up the constant refuse from dead wildlife that otherwise would present health hazards. In one infamous case, the impact of the loss of birds was starkly obvious. Three primary species of vultures that patrol India and Pakistan died off suddenly from an anti-inflammatory drug used on cows that the vultures in turn ingested. That left cow carcasses spread all over the country, which vultures would normally have consumed. This led to rabies and other diseases that killed thousands of people with an overall economic impact estimated at $34 billion. Erin Katzner, director of global engagement for the Boise-based Peregrine Fund, which helped search for solutions, says the case reflects how little we understand the balance of nature and how careful we should be about undermining it. "Until one of these birds starts to go extinct," she says, "we don't know what the consequences can be."

Over the past two years, we heard many leading ornithologists talk about what birds mean to us—and what our responsibilities should be to them. John Fitzpatrick has spent his career finding fresh ways to explain the sanctity of birds. "They're among nature's masterpieces," he says. "They speak to our hearts." He told us about the latest presentation he's developed that compares the *Mona Lisa* to the last few endangered birds in Argentina called Hooded Grebes. He starts the talk with a video showing the birds in their exotic, boisterous mating dance, and zeroes in on their bright red eyes and reddish brown crest that rises to a point at the top of their heads. Then he switches from the bird, whose history dates back 30 million years, to a photo of Leonardo da Vinci's painting.

"We spend millions of dollars protecting artwork like this, a couple of hundred years old," he says. "But what we're about to lose in Argentina is infinitely more complex with a vastly more elaborate history. Why do we have to convince people that it's worth saving this? Is it worth investing in the great works of art in the world? Of course, it is. So are these birds—not just because we depend on them, but because this is a feature of the living earth, and we have a chance to be alive during its time. The idea of losing it is preposterous."

Coming Home

After traveling the equivalent of the circumference of the globe and conducting some three hundred interviews with people across the world of birds, we parked our Airstream and started the unloading stage. It took just a few days to put away the dirty laundry, camping supplies, leftover canned goods, books, and boxes of interview transcripts. But unpacking the complicated, slow-motion drama of the North American birdscape, that took months.

We returned home inspired by the work under way to save birds. We met folks who ruin their knees scrambling along dangerous cliffs, agonize over algorithms, confront adversaries at gunpoint, and sometimes get their eyebrows singed off. They welcomed us into their lives for days at a time and shared their hopes, frustrations, and determination. Taken together, their experiences help make the case for birds—not only as nature's workhorses and cultural icons, but as living bellwethers of the environment at a pivotal time. As the National Audubon Society's Elizabeth Gray put it, "Birds are telling us what is happening with the planet, and they're telling us what we need to do."

This is without question a rare juncture. The past few years have brought a far richer understanding of what birds contribute and what they're up against in an environment remade for human needs. The latest research and technology provide the equivalent of an MRI of the continent's avian landscape. Massive bioacoustic sweeps and precision-tracking tools paint detailed portraits of individual species down to the neighborhood level. These advances provide what can be roadmaps for shoring up bird populations, identifying where habitat needs are greatest and hazards most severe. New conservation concepts show how birds and people can coexist on ranches, farms, and forests, in suburbs, military bases, and cities—if we're willing to make room.

Every time bird populations have been threatened over more than a century, from the feathered hat craze up through the collapse of ducks and the decimation of the Bald Eagle, the country has come to the rescue. Congress passed the Migratory Bird Treaty Act in 1918, the Endangered Species Act in 1973, the North American Wetlands Conservation Act in 1989, and other legislation, all of which set world standards for bird protection. Lawmakers taxed guns and ammunition to fund science, conservation, and refuges to manage the broad recovery of ducks and geese. All fifty states have wildlife agencies that help federal agencies meet the nation's mandate to protect wildlife. Our laws plainly state that this nation will watch over birds as part of our natural resources.

This time around, though, the challenges of delivering on these standards are more complex and the solutions more costly—and we as a country have yet to commit to living up to these principles. The system built for protecting wildlife is falling steadily further behind. Funding is inadequate to keep up with even determining which species are in need of help. Too much of the recovery effort and the funds that go with it are focused on the last-minute safety net for troubled species, and not enough on birds before they reach the crisis stage. Almost no effort is addressing one of the most startling trends, that the biggest losses are now coming to such common birds as sparrows, blackbirds, larks, and finches. "We're not saying these birds are going extinct," says Ken Rosenberg, the now retired lead author of the Three Billion Bird report. "We're saying we should be paying attention to them far earlier when we can make a difference."

The primary federal agency responsible for birds has yet to rise to the occasion. Instead, the U.S. Fish and Wildlife Service seems content to manage the declines. It's been unwilling to take significant steps to confront the loss of a third of North America's birds, or even to address some of the most pressing contributors that so obviously merit action. At a time when wind turbines are spreading rapidly, for instance, federal agencies don't have a coherent policy for avoiding the collisions caused by placing turbines in migratory pathways. The agency's most successful conservation venture in its history, the revival of the Bald Eagle, faces threats from poisoning caused by lead bullets. A massive study released in 2022 concluded that more than half of the 1,210 Bald and Golden Eagles tested in thirty-eight states had high levels of lead in their systems. This ammunition was long ago prohibited for duck hunting, since people eat the game birds. But the agency has declined to use its authority to protect our nation's symbol from an excruciating death that comes to these magnificent birds from consuming carrion left in the field contaminated with lead.

These failures symbolize the inadequate current policies of the Fish and Wildlife Service. By keeping a low profile on these issues, the agency is missing an opportunity to provide a voice on behalf of birds at a crucial time. With research, technology, and knowledge for understanding birds reaching new heights, the agency should be making the most of these advances. Even insiders who've devoted their careers to the agency say they're frustrated with what they're seeing. Don Barry, the former chief counsel at U.S. Fish and Wildlife, who helped shape the agency's laws and enforcement for years, is direct about it: "Unfortunately over the last twenty years, the Fish and Wildlife Service has grown increasingly timid." Dan Ashe, director of the agency for almost a decade during the Obama administration, says, "They just don't have the depth of field that's necessary to deal with the challenges they're facing."

The system is in disarray, but there's no strategic discussion taking place in Congress on what to do about it. The partisan divide has stalled any negotiations over whether the Endangered Species Act should be updated to reflect this modern landscape. The legislation itself hasn't been reauthorized—a routine matter for ongoing policies—since 1992. Lawmakers have been unwilling to even take up the debate for fear one side or the other will lose ground. Democrats are afraid Republicans will try to water down its protections, and Republicans don't want to risk seeing the powers expanded. This leaves the fate of troubled birds hanging. Some are left to steadily decline while waiting for years for the basic research to determine their status. On average, rulings on whether to add species to the ESA list take twelve years, a time frame that undermines the only last-resort recovery option birds currently have. More than fifty species of wildlife and plants have gone extinct waiting for the Wildlife Service reviews in recent decades, according to a protest letter from 140 wildlife agencies sent to the Fish and Wildlife agency in April of 2022 calling for reforms.

On a more hopeful front, a cross section of nonprofits, research centers, philanthropies, and a few government agencies are putting together strategies that do confront this crisis. Nonprofits from Audubon to Cornell, ABC to Ducks Unlimited are starting to join forces on key projects to protect birds in the United States and Latin America. Growing alarm over climate change is helping raise foundation funds for land purchases and preservation aimed at climate, but that has the added benefit of providing vast new stretches of open land for birds.

It's not hard to envision what specific actions could—and should— be taken. We found workable remedies spread out around the country like pieces of a puzzle. The success of the continent's combined private-public policies for ducks and geese shows the power of targeted habitat restoration, adequate funding, and grassroots support. Despite extensive development along coasts and waterways, game bird populations are up more than 50 percent—a stunning achievement that ought to be a model for the overall strategy for birds. Meanwhile, the Department of Defense has proven that practical research and aggressive conservation can safeguard birds in even the most hazardous settings. If the U.S. Army can manage to protect endangered birds amid bombing ranges and runways while training for combat, why aren't these lessons put into place by other state and federal agencies that oversee wildlife habitats?

These examples illustrate something else as well: Taking simple steps to care for natural resources delivers benefits that go beyond the restoration of these species alone. Ryan Orndorff, director of natural resources for the Department of Defense in Washington, D.C., says the country's armed forces are much better off with the wildlife conservation measures that are now commonplace in the military. "It's good for the species, and it's good for our overall mission," he says. "It's in the nation's best interest that we have a prepared armed forces. It's also in the nation's best interest that these imperiled species remain in existence."

Businesses are discovering that birds can serve as guides for helping restore altered environments. The early success of companies like Walmart and Nespresso in using birds as indicators is getting attention from a growing list of curious competitors who don't want to be left behind. Researchers are seeing a pivotal role for agriculture in helping provide a layer of conservation that links together public lands. "Conservation is going to be extraordinarily dependent on working farms and working ranches," says Hilary Swain, the director of Florida's Archbold Biological Station, which operates the research-based Buck Island Ranch. "Our ability to succeed without them will be greatly limited."

The United States is a wealthy, generous country—but balks when it comes to fully funding bird conservation. This is despite the benefits we now know come in the health of the environment, productivity of the land, and sustaining the balance of nature. We'll spend $40 billion for a repeat trip to the moon. We invest $200 billion annually in state and federal highways, and we spend $80 billion a year on state jails and federal prisons. But we're reluctant to spend a small fraction of those amounts to keep up with the needs of conservation and research. One of the most comprehensive looks at these costs found that the expense of halting the extinction of birds and safeguarding important areas for biodiversity in the United States would come to $1.23 billion a year—about 20 percent of what this country spends on soft drinks. The funding needs for conservation are growing, but it's an enormously good deal compared to the long-term financial consequences of not taking action.

A Wing and a Prayer

The contradiction in conservation is that the public widely supports the idea of protecting wildlife, according to numerous opinion polls. But there's a divergence of views on what exactly that should look like. The country's political divide figures into how people see tactics for conservation: Some argue that caring for wildlife should be voluntary, left up to individuals; others think state and federal agencies need to take a stronger stand. We found instances of clear success from both of these approaches: The ranchers we met in Wyoming see their voluntary Sage Grouse Initiative as an example of what can be done on private land. The success of rescue missions driven by government agencies for the California Spotted Owl, Grasshopper Sparrow, Whooping Crane, California Condor, and Florida Scrub-Jay —it's a long list—illustrate what can be done with the support of the law and at least some ongoing funding. The Department of Agriculture's conservation spending for more than one hundred birds shows how applying incentives in the nation's breadbasket can help ease the impacts of agriculture and help protect birds in the middle of working farms and ranches.

The fact is an assortment of tactics are needed to halt the further loss of the hemisphere's birds. The various perspectives don't have to be at odds; they can be complementary and designed to fit with regional viewpoints. We do need to follow the federal laws that have been at the heart of the country's conservation successes for fifty years, and the Endangered Species Act must be funded at a level that will provide the safety net that has never been more warranted. But those steps will not succeed unless private property owners are encouraged to be part of the solution. Many of the most meaningful projects are fueled by local people caring for the land where they live, work, and play. "In the end," says Archbold Station's Swain, "all conservation is local."

If there's an overarching lesson we've learned in our own travels, it is this: The work of safeguarding birds and the environment we share with them cannot be left to a handful of scientists and wildlife agencies alone. Conservation measures need to be shared by a far wider segment of people, organizations, and businesses. The benefits of making room for birds are similar whether it's the 182,000 acres of the Army's Fort Benning or a quarter acre backyard.

Thanks to technology, anyone can grab a front-row seat—call it a bird's-eye view—to watch this fascinating parallel world that was once hidden away. We can follow migration paths with smartphone apps and signals beaming down from space. We can practically move into the nests with eagles, herons, condors, and hawks with the help of hundreds of bird cams. We can contribute data from the species we see that adds to the storehouse of knowledge that helps birds. There are all kinds of ways each of us can support this cause, many of which are listed in the Afterword that follows.

When we first began this project, we went looking for a metaphor that captures the struggle to save birds. We settled on the title "A Wing and a Prayer" after we learned the story behind the phrase. It dates back to World War II, when American pilot Hugh G. Ashcraft Jr. returned from a bombing raid over Germany with the plane's tail in shreds and one engine in flames. As the bomber lost altitude approaching the coast of England, Ashcraft suggested his crew might want to say their prayers and prepare for the worst.

But the flight did reach home, and the news of its good fortune took on a life of its own. The actual saying was a creation of Hollywood in a 1942 movie called *Flying Tigers*. The star was John Wayne, who runs into headquarters looking for news of the damaged plane as it approaches its home base. That's when a clerk delivers the line: "It's coming in on a wing and a prayer." Something about the power and poetry of those words instantly resonated with people. The phrase has since gone on to inspire poems, books, sermons, popular songs, and more movies, all built on the idea of going against long odds in a valiant cause.

Many birds are facing long odds, and the cause is a compelling one that is part of the environmental story of our time. The more we learn about what's at stake with the future of this treasury of nature, the more hopeful we are this country will do what should be done. The birds don't need all that much from us, and what's needed will have as many benefits for people as they will for birds. Pete Marra has a knack for putting it plainly. "If we can save birds," the Georgetown scientist says, "we can save ourselves."

The Three Billion Bird study stripped all mystery from the troubled state of the hemisphere's birdscape. There's still time to respond, but that time is now. It's clear what steps are making a difference and what will help avoid another half century like the last one. Halting the collapse of our birds will not be easy. But as the scores of researchers, birders, wildlife experts, hunters, and philanthropists are proving every day, a turnaround is within reach if we'll listen to what the birds are telling us.

The male Cedar Waxwing passes an insect to its slightly smaller female counterpart as part of the bird's courting ritual.

HOW YOU CAN HELP

From Beverly:

Wherever you are . . .

If you're wondering how to help birds, there's something for everyone and at every level. The smorgasbord of remedies includes anything from planting a native bush and drinking shade-grown coffee to buying the right gifts and engaging in a bit of sly party conversation. You can also donate to the cause through dozens of different organizations.

There's the real power to make an impact if we all start somewhere. And if you'd like to dive deeper, specific details and next steps for each suggestion can be found on our website, www.FlyingLessons.US.

One logical place to start is to help keep birds alive. There are two "situations" that top most lists for how to make the greatest impact. Plus they have feel-good potential attached.

SITUATION #1: One sunny morning in May years ago in downtown Washington, D.C., we heard a thump against our balcony glass door. It was a Yellow Warbler now motionless on the concrete floor. Until that day we didn't realize that up to a billion birds die every year in the United States and Canada crashing into glass. The birds perceive the reflection as just more airspace.

Roughly half of these strikes are window collisions occurring at people's homes. Preventing them can be as simple as putting decals or screens on our glass doors and windows. The American Bird Conservancy's website lists twenty-nine readily available products that you can order. Here's the link: https://abcbirds.org/glass-collisions/products-database/.

The other half involve collisions with high-rise buildings. Many cities have "lights out nights" because that's when migrating birds tend to fly. They're attracted by the lights and crash into buildings. If your city doesn't do "Lights Out for Birds" perhaps you can start a letter-writing campaign to city officials. For these details, check our Flying Lessons website.

As for our stunned Yellow Warbler on the balcony, the lucky bird got back on his feet and eventually flew away.

SITUATION #2: Sad but true: Aside from habitat loss, cats are the number one reason for bird deaths that people can prevent. One day, I got a firsthand look when a friend and I were lazing around on her lawn. Her cat slinks up and proudly drops a dead sparrow at her feet. Regardless of how well fed, cats are born to hunt. They often share their bounty, so when a cat brings you a dead animal, they consider you a part of their family. So adorable! Until you learn that outdoor cats are estimated to kill more than 2.6 billion birds annually in the United States and Canada. So, if you have a housecat, keep it indoors. It's healthier for them, too. Or you can create an outdoor "catio" enclosure so your cat has a safe way to enjoy the fresh air. Those clever cats can also be trained to walk on a leash.

DON'T HAVE A CAT? GOSSIP ABOUT THEM: You can still help the cause by bringing this up in conversations with friends. The sheer numbers are such a shocker you'll have no trouble getting their attention. While you're at it, give equal time to the canines by passing along this slogan: "Bird-friendly beaches have dogs on leashes!" Summer is beach time, and it's also when migrating birds such as plovers, skimmers, and terns lay their eggs in the sand. One of the biggest challenges shorebirds face is unleashed dogs that harass them and trample nests. Dogs will sometimes even eat the chicks. Spread the word with friends and on social media and help save birds.

ENJOY BIRDS WITH A FAVORITE CHILD: When it comes to entertainment value, sometimes our feathered friends might as well be puppies. As children gradually learn how to watch birds and understand their antics, they're getting a unique peek into a different world. And who knows, perhaps you'll be inspiring a future conservationist.

Start slowly by pointing out birds on walks, looking at beautiful photos, or helping the child identify which songs go with which birds. You can gradually do more as the child grows—everything from hiking in a nature preserve, using a children's bird guide together, taking photos, or signing them up for nature day camp. Audubon has a book, *Audubon Birding Adventures for Kids*, that details all sorts of other activities. If you really get inspired or know a bird-friendly educator, the Cornell Lab has free K–12 science lessons that anyone can download.

EAT PANCAKES WITH BIRD-FRIENDLY SYRUP: Organic maple syrup with the "Produced in Bird-friendly Habitats" certification promotes biodiversity in sugarbushes so songbirds can continue to forage, find cover, and raise their young. There are links to a bunch of online vendors on National Audubon's website at www.audubon.org/maple.

Other bird-friendly foods such as beef, chocolate, and rice are inching toward the marketplace (the challenges are similar to coffee, see below), so stay on the lookout. Currently most of these are independently produced and sold in local specialty stores or online by the producers. Audubon is hoping its Bird-Friendly Beef brand—with a green certification label featuring a cow and meadowlark side by side—will become widespread. Its website includes a map of where you can find groceries, butcher shops, and online suppliers around the country. https://www.audubon.org/conservation/ranching.

DRINK COFFEE THAT'S GOOD FOR BIRDS: This is a bit complicated, but here's why it matters: Three quarters of the world's coffee farms grow their plants in the sun, destroying forests that shelter and feed birds and other wildlife in some of the richest territories in the world for biodiversity. However, shade-grown coffee (the traditional growing method), protects forest canopies migratory birds use to survive the winter. The Smithsonian Migratory Bird Center started a "Bird Friendly" certification that is the gold standard for shade-grown, organic coffee. Unfortunately, the program struggles against multiple forces, including how costly it is for small farmers to restore degraded farmland.

Currently, the only major grocer that sells Bird Friendly coffee is Whole Foods. It's also available on Amazon and at a number of small producers around the country. It can be hard to find bird-friendly coffee in many places, but keep in mind that you're still helping birds by buying shade-grown or organic coffee that's more available than the full bird-friendly brands. Small steps do reap rewards.

SHOP—FOR GIFTS OR FOR YOURSELF: There's just so much to buy that gives back to conservation it makes me giddy. I love shopping, and a little retail therapy on behalf of birds is irresistible. Who cares if half of what

I wear is covered with wings? Bird Collective sells my new favorites, particularly sweatshirts, hats, and T-shirts with artistically arranged, exquisitely embroidered songbirds. This collection is a collaboration with Audubon's Birdsong Project. Bird Collective is a small, women-owned business that donates a portion of its profits to conservation groups, including the American Bird Conservancy, HawkWatch International, and the Grassland Bird Trust.

Gift shops at National Wildlife Refuges, nature centers, the Cornell Lab, and zoos are terrific places to shop and benefit birds and other wildlife. Audubon's online Marketplace has logo items like backpacks and totes, but also bird-shaped sun catchers and gardening tools. Audubon also gets funding by licensing its brand for products sold by other vendors, including books, games, puzzles, and birdseed.

BUY BINOCULARS, A GIFT THAT KEEPS ON GIVING: When looking at birds through binoculars, a blur of feathers becomes a magnificent canvas of colors, textures, and patterns. No two birds are alike, and the subtle differences put nature into a new perspective. The joy you'll get from binoculars is worth every penny. Once Anders and I watched a female Cedar Waxwing lean toward the male as he fed her insects one by one. It was just the sweetest moment, as you can see by the photo at the top of this section.

So how does this help birds? I would argue it's difficult to care about birds you never see, and that's why buying yourself some binoculars is ultimately good for birds. In general, the more you watch any bird, the more you're able to appreciate it. And the more you appreciate birds, the more likely you are to care about what happens to them. But you really do need to get up close.

If you're planning to buy binoculars, visit the Birding Basics section of our Flying Lessons website for suggestions. And be sure to check out our tips and tricks on how to use binoculars specifically to see birds. (It's not as obvious as you'd think, and they rarely come with instructions.)

WATCH A BIRD CAM: If it's impossible to see birds in person, the next best thing is watching a bird cam. These are close-up cameras focused inside a nest, and they broadcast live footage nonstop online during breeding season. Some friends in California got so enthralled they moved an old computer into the kitchen and continuously broadcast the bird cam channel. That way they could easily keep track of what quickly became "their" birds. So pick a cam to watch and tell your friends. A quick internet search is the best way to find one.

SPREAD THE MESSAGE: When you fall in love with birds, recruit your friends. One of the best ways to lure people into birding is to show them how to use the Cornell Lab's free Merlin Bird ID smartphone app. We've found Merlin is one of the easiest ways to learn about birds and know what's flying around you. Even people who aren't particularly into birds enjoy watching Merlin's Sound ID instantly identify which birds are singing in real time. It's like a delightful party trick.

We're always looking for opportunities to help our neighbors and friends get more interested in birds. We invite them to join us on walks, share our binoculars, and point them to birding hotspots we know they'll enjoy on their own. Ultimately your friends will thank you—ours certainly have. You're giving them the gift of a rich and rewarding pastime, and the chance to have a true relationship with birds that will last a lifetime.

INSTALL A BIRD FEEDER: Feeders lure birds to your home so you can get to know them, and they help birds survive the winter when food is scarce. Feeders come in so many styles and prices it can be confusing. If you want a quick, sure pick for most types of birds, a "Wirecutter" review team from *The New York Times* recommends the Droll Yankees 18-inch Onyx Mixed Seed Tube Bird Feeder with Removable Base. The Onyx feeder has superior construction, and its bottom can be removed for easy cleaning.

If you want an up-close visit with birds at your feeder, sit as near to it as you can without intimidating the birds. Wear clothes that blend in with the surroundings (no whites or brights), stay very still, be very quiet, and just wait for the show. It's worth being patient. Focus on a chickadee for example, and you'll see the saucy tilt of the head and what I like to think are amusing expressions. See if you can figure out which cardinals are mated pairs (they'll often visit together and take turns), or which hummingbird is the dominant male (he chases away the others).

If you want to attract hummingbirds, there are feeders for less than $10. No need to buy red food. Just mix 1 part sugar to 4 parts water until the sugar dissolves. (If you heat it, cool the syrup completely before adding it to the feeder.) Refrigerate any extra until ready to use.

BIRDHOUSE BASICS: Birdhouses are widely used as decorative yard accessories, but the wrong type of house can actually harm birds. What most birds need are boxes made of unpainted wood that's stained with a natural wood preservative such as linseed oil. No perch! Birds don't need them, and predators will use them.

There are about three dozen types of cavity-nesting birds that will typically use houses, including bluebirds, chickadees, titmice, robins, and wrens. These are also called nest boxes, and their requirements for size, shape, etc., vary by species and by location, so you'll need to do a bit of research. State and local Audubon chapter websites usually offer specifics for birds in your area. Staff at a local hardware, garden store, or specialty shop like Wild Birds Unlimited can usually help.

Take It Up a Notch: More Ways to Have Impact

SPEAK UP: You can just click an online form to let Congress know you care about legislation benefiting birds and other wildlife. From the websites of both Audubon and American Bird Conservancy, you can fill out a quick form that will be automatically sent to your senator and representatives. These clicks do add up. In 2021, Audubon facilitated more than 170,000 people in contacting decision makers 1,085,000 times about upcoming legislation benefiting birds and wildlife. I recently urged my representatives to vote in favor of the Saving America's Pollinators Act, which would prohibit the use of insecticides that are harmful to birds and many of the insects that birds eat. And my representative emailed me back, so someone in her office took note.

You can find the Audubon link at https://www.audubon.org/take action. The link for American Bird Conservancy is https://abcbirds .org/get-involved/take-action/.

SHARE WHAT YOU SEE: Join a Cornell Lab project such as eBird or Project FeederWatch to record your bird observations. Your contributions provide valuable information that tells scientists where birds are thriving— and where they need help. You can download the eBird app for free from your smartphone app store. The Cornell Lab also offers a free course on how to use eBird via its website.

PUT IN NATIVE PLANTS: Making a difference in your yard is as easy as providing food, water, and places for birds to hide and build nests. The National Wildlife Federation and its twenty-four state affiliates started its Garden for Wildlife program in 1973 with the goal of empowering people to turn their own small piece of the earth into thriving habitats. Its website has lots of home how-to resources at this link: https://www .nwf.org/Garden-for-Wildlife/Create/At-Home.

Different types of native plants play different roles. Caterpillars, for example, are a critical food source for over 96 percent of songbirds. For a brood of nestlings, a Carolina Chickadee needs up to 9,000 caterpillars that live in native oaks and other trees. To find out which plants in your climate zone support butterflies, moths, and birds, you can plug in your zip code on a tool called the Native Plant Finder on the National Wildlife Federation's website.

You might need to make an effort to find propagated native plants. Garden for Wildlife also offers another chance to go shopping and benefit birds: The site's plant store offers collections of native plants designed for thirty-six states in the Northeast, Mid-Atlantic, Southeast, and Midwest. You can also purchase gift cards.

HERE'S A PLUG FOR WOODPECKERS: Leave dead trees where they lie. When they're not a threat to the surroundings, dead trees provide homes for woodpeckers. Without their leaves, dead trees look like natural outdoor sculptures.

AVOID PESTICIDES: The continent's most widely used insecticides, the neo-nicotinoids (or neonics), can be lethal to birds, bees, and to the insects that birds rely on for food. At home, avoid using common weed killers, like Roundup, which contain glyphosate. There are other neonic insecticides that are widely used on farms to pre-treat crops, enabling them to collect in high concentrations in surface water that birds drink and on the surface of seeds that birds eat. (The evidence is so far inconclusive to what extent neonics harm humans, but many European countries have banned them. It's worth considering birds are still our canaries in the coal mine.) The best ways to help birds and humans are to write to your congressional representatives and vote with your wallet by buying organic food.

PROTECT SEABIRDS FROM PLASTICS: There are lots of reasons to cut down on plastics, but if you avoid them near the beach, seabirds will benefit. There are eighty varieties of seabirds that are especially vulnerable be-cause they mistake plastic for food. Even toothbrushes and cigarette lighters have been found in the stomachs of dead albatrosses. If you do use plastic, take it home to recycle. Next time I'm at the beach, I'm going to bring a garbage bag and pick up plastic along with shells on walks. For those who live near the beach, it could be fun to organize a beach cleanup outing with friends.

JOIN ONE OF THE BIRD NONPROFITS: Let's say you want to help birds right this very minute, while you're motivated, and it must be easy, quick, and guaranteed to have an impact. I suggest joining a conservation organization focused specifically on birds. If you already belong to one, join a new group or give someone a gift membership. The primary groups, all top rated by Charity Navigator, are the National Audubon Society and its local chapters, the American Bird Conservancy, the Bird Conservancy of the Rockies, and the Cornell Lab of Ornithology.

You can join online, dues are tax-deductible donations and vary from $20 to $45, depending on the group. Some organizations include a subscription to their beautiful bird magazines.

There are also groups that focus on specific birds, including the Birds of Prey Foundation, International Crane Foundation, American Eagle Foundation, and the North American Bluebird Society. If you're concerned about waterfowl and wetlands, Ducks Unlimited is the obvious choice. Then there are other nonprofits that conserve nature more broadly while also providing essential help for birds. Well known in this category are The Nature Conservancy, Center for Biodiversity, the National Wildlife Refuge Association, and the National Wildlife Federation.

There's power in numbers, and if everyone joins an organization, we can make a significant difference in the day-to-day campaign to save our vanishing birds.

A Final Postscipt: How to Be a Conservation Birder

From Beverly and Anders:

We first crossed paths with Sharon Pitcairn Forsyth by chance. Beverly was peering through binoculars at the entrance of Rock Creek Park in Washington, D.C., on a winter afternoon some years ago, and Sharon was having an intense phone conversation nearby. But she spotted us for birders, covered over her phone, and took a moment to relay some exciting news: About fifty yards away, a rare yellow-and-brown bird was rustling in the leaves. "See where that person is taking photos?" she told us. "That's where it is."

This bird, a Dickcissel, was one of our first rare finds. It's a grassland bird that breeds in the central U.S., begins its migration by the end of August, and is usually ensconced mostly in Venezuela by November. But here it was in chilly D.C. in January, making the Dickcissel a local curiosity because it veered from its migration route and settled where it wasn't supposed to be.

We wouldn't learn whom to thank for the tipoff or officially meet her for several weeks, but that happenstance encounter was the start of something magical for us. Becoming a better birder can be a slow and confusing process. Guidebooks and online videos will get you started. But if you want to fast-forward your skills, the best way is to get help from other experienced birders.

Unfortunately, you can't always count on that. While it may look like a leisurely stroll through the woods, birding is often a competitive sport with players intent chiefly on adding unseen birds to their life list. At an even higher level are the twitchers, the term reserved for birders who travel the country and sometimes the world to see rarities. In the beginning we didn't realize birding had a hierarchy, a set of do's and don'ts, or that veterans don't always welcome the uninitiated.

Take the morning early in our birding life when we parked in a Northern Virginia nature preserve alongside a group of "real" birders—folks wearing khaki vests, bucket hats, and hiking boots. They were finished for the day and putting cannon-size cameras into their trunk. Grinning and full of questions, Beverly—wearing a hot-pink cap and matching T-shirt along with binoculars—hurried out of our car. As soon as they saw us, the birders jumped into theirs as if we might be planning an attack. In their defense, we sort of were.

And then there was the time she got banned from a rare bird

alert in North Carolina. This is a system where, via a phone app, area birders—local twitchers in effect—ping the group when they see a rare bird so others can go find it, too. We were new to town and wanted to know where a certain endangered woodpecker might be. Who better to ask than the area's best birders? We posted, and sure enough, some great advice started rolling in.

However, a few minutes later we got a nasty message from the group's moderator: This was a private platform for rare sightings only. Okay, good to know. Several weeks later, being careful to stay on topic, we put up a post asking for details about a current area sighting. Now we'd broken the rules twice and apparently annoyed the other birders to boot. And that's when the moderator rapped our knuckles and kicked us off the message group permanently. So let this be a lesson: There do exist birders for whom two questions are one too many.

Sharon, however, is a birder of a completely different species. She's been the unofficial social director of D.C.'s birding network, and she chose to see potential in our keen enthusiasm. She's definitely a twitcher, but that doesn't keep her from helping beginners. She recognizes that expanding the tribe is likely to be good for birds.

"What are your favorite types of birds?" she asked the first time we got together for a spring migration walk. "Warblers," Beverly said. Sharon grabbed the challenge. "Okay, which ones haven't you seen?" Before long we all were off in search of Worm-eating Warblers, tiny earth-toned migratory birds that prefer dense leafy forests and are therefore difficult to find. But Sharon knew just where to go. When we arrived, voilà, two delightful Worm-eating Warblers were soon flitting about. It seemed to us as if she pulled those birds right out of her hat.

Sharon's claim to fame could be her penchant for hospitable birding, but it turns out she's something that's even more rare. The child of conservation-minded parents and the wife of scientist and author Adrian Forsyth, whose career focus is preserving the Amazon, Sharon developed a passion for biodiversity long before the concept became commonplace. She's what we've come to think of as a "Conservation Birder." That's someone who loves saving birds and their habitat as much as they do watching them.

You'd think the woods would be full of birders interested in actively helping our feathered treasures. Unfortunately that's not the case. George Fenwick, the American Bird Conservancy's first director, explains in the foreword of the organization's *Guide to Bird Conservation* that birders tend to be people who prefer to walk in the woods rather than advocate for the interests of birds. "I have read or been told time after time by birders that they want their hobby to be just that—a relaxing pastime, and not a cause for which to fight or pay," he writes. "This means that only a fraction of birders wish to play a role in protecting the resource they enjoy."

Furthermore, Fenwick points out that the mere fact that there are 50 million people who like to look at birds will not be enough to convince society to cough up resources for bird conservation. In other words, we can't just stand around hiding behind our binoculars. "Instead, we will succeed only if we boldly speak up," Fenwick says.

The bird organizations are all struggling with how to encourage their members to be more active to push for progress and how to widen the circle in birding from what has traditionally been a mostly older and largely white following. The Cornell Lab's Ian Owens sees the goal as "building a movement for nature." That's the same phrase that Audubon's chief, Elizabeth Gray, uses for what's needed—and the logical foundation for that is with birders. "We're helping to build a movement to engage those millions of birders to drive conservation decisions," she says.

Now is certainly the time to do just that. But we suspect there's a deeper obstacle: For birders, engaging with such severe losses and the sorry state of so many habitats can feel overwhelming, stressful, depressing. Just pick an awful word, and that's where we are. So what if birding starts to feel like watching your friends face a firing squad—where's the fun and relaxation in that?

Well, it's an interesting paradox, but one that has to be overcome if we're all to have birds to watch in the future. The bird organizations are critical to uniting people in this cause. But just as important is the work of individuals willing to bring along the newcomers who are now finding their way to birding by the droves. Conservation Birders like Sharon are a model for how it's done.

In the end, our friend and mentor did more for us—and for the birds—than welcome us into the tribe. Bit by bit and bird by bird, she guided us down the path to becoming Conservation Birders, too. It's fair to say if she hadn't taken us under her wing, we'd never have written this book. And what we ultimately learned in the process is that as birders, we do need to grow our ranks. We also need to be far more welcoming and willing to step up, do our part to care for what we love, and nudge others do the same. Be sure to walk softly, smile broadly, and figure out which birds those newbies might like to see.

Acknowledgments

Toward the end of a week of solid rain, parked in a dilapidated campground in Louisiana with water creeping up our Airstream's hubcaps, we looked at each other in silence. "Do you think it's too late to get out of this?" Anders finally asked. It was still early in our journey, but far enough along to understand the dimensions of what lay ahead. The only reason we now have the chance to share our thanks and appreciation is because of the long list of friends, family, a few kind strangers, and the many subjects of these stories who helped us find our way.

The rain eventually stopped that week in Louisiana and we escaped down the road to the Texas coast to find the most inspiring collection of warblers we've ever seen. We've always thought of these tiny, colorful migrants as feathered gifts—and this batch knew just when to make an appearance. Today, we wonder how we could have been so lucky to travel all those thousands of miles almost entirely under sunny skies, along mostly passable roads, and surrounded by birds.

We don't pretend to be experts, so we cannot overstate the value of the hours of patient, thoughtful guidance so many people gave us over the past two years. The Cornell Lab of Ornithology opened every part of the center to us, right down to the row of three Ivory-billed Woodpeckers in its specimen collection. The staff and leadership of the

American Bird Conservancy not only put up with hours of interviews, but let us tag along on an expedition through Ecuador. The National Audubon Society, Nature Conservancy, Archbold Biological Station, White Oak Conservation Center, Ducks Unlimited, Tall Timbers, the Fort Benning U.S. Army Base, the Department of Defense's Strategic Environmental Research and Development Program, U.S. Fish and Wildlife Service, U.S. Forest Service, U.S. Department of Agriculture, Point Blue Conservation Science, Peregrine Fund, San Diego Zoo Wildlife Alliance, U.S. Geological Survey's Bird Banding Laboratory, Rare Species Conservatory Foundation, Project Principalis, National Aviary, Bobolink Foundation, Jocotoco Foundation, Jack Jeffrey's guiding and photography, BirdLife International, Motus Wildlife Tracking System, the ICARUS project, Defenders of Wildlife, and Center for Biological Diversity all shared their time and wisdom with us. We know our lengthy interviews, repeat conversations, and follow-up questions must have gotten wearisome long before we asked the last round.

The true guiding light on this project was Mindy Marqués, our editor at Simon & Schuster. She saw the potential from the start and nursed this project from a sketch to reality. Every time she popped up on our Zoom calls, she delivered the encouragement and insights we needed. We couldn't have been luckier to have the cover design land on the desk of Natalia Olbinski, who happens to be a birder; the photography, editing, and endless publication questions fall to Hana Park; and the superb editing of Fred Chase. A special thanks to Pete Cross, who not only took photos of us with our Airstream during visits to Florida and shot the back cover photo, but is such an inspiration and guide on all things photography.

We'll always treasure the day, sitting in the back row of a Gridiron Dinner music rehearsal, when our friend Robin Sproul asked about our birding life. Two years later, she hadn't forgotten and called to suggest writing a book. Robin, now an executive vice president at the Javelin literary agency, along with Javelin founders Matt Latimer and Keith

Urbahn, saw the potential and shaped our idea. When it came time to kick titles around, Matt—with his impeccable sense of story—was the one who came up with "A Wing and a Prayer."

Our circle of friends and family who began reading long before this project was cooked had immeasurable impact on these pages. At the top of the list are Liza and Bill Bennett, who talked us through the project from rough ideas, through early drafts to final touches. We're so thankful for the guidance from Ed Hatcher, Angie Cannon, Robbi Farrell, Stephen Gyllenhaal, Joby Warrick, Joe Starita, Sharon Pitcairn Forsyth, Adrian Forsyth, Mike McDonald, Kathy Hart, Phyllis Azar, and Ramona DuBose.

We ended up pitching camp in every imaginable setting, from the edge of San Diego Bay to within feet of Currituck Sound on North Carolina's Knotts Island. We were most thankful, though, for the driveways we landed on and the ensuing hospitality from Louanne and Fred Bisel, Stephen and Kathleen Gyllenhaal, Jeff and Bev Partyka, Phil and Rochelle Ward, Pete Cross and Christie Evans, and the Land Yacht Harbor Airstream Park community.

Finally, we're forever grateful for the love and support of our children, Sam and Grey, who met us on stops along the way, put up with our endless talk about birds, and provided a steady blend of encouragement and levity that made all the difference.

Notes

Introduction: What the Birds Are Telling Us

2 *In the past fifty years, nearly a third of the bird population:* Kenneth V. Rosenberg et al., "Decline of the North American Avifauna," *Science*, Sept. 19, 2019, https://www.science.org/doi/10.1126/science.aaw1313.

3 *As one veteran biologist:* Interview with John Doresky, Columbus, Ga., March 23, 2021.

3 *One nonprofit has landed a multimillion-dollar donation:* Interview with Ben Novak, Brevard, N.C., Nov. 29, 2021.

4 *"This is a crisis":* Interview with Nadine Lamberski, San Diego, Calif., May 10, 2021.

4 *Birds consume an estimated:* Martin Nyffeler et al., "Insectivorous Birds Consume an Estimated 400 to 500 Million Tons of Prey Annually," *Science of Nature*, July 9, 2018, https://www.researchgate.net/publica tion/326272395_Insectivorous_birds_consume_an_estimated_400-500 _million_tons_of_prey_annually.

4 *Researchers have discovered that watching birds:* Julia John, "Seeing Birds Can Help People Destress," The Wildlife Society, March 23, 2017, https://wildlife.org/seeing-birds-can-help-people-de-stress/.

5 *They are among its original inhabitants:* Sara Goudarzi, "Modern Birds Existed Before Dinosaurs Die-off," *National Geographic*, Feb. 8, 2008.

5 *As recently as 150 years ago:* Interview with Ben Novak, Brevard, N.C., Nov. 29, 2021.

5 *That changed in 2019 when a group of researchers announced":* Based on interviews with authors of the Three Billion Bird report in 2021 and 2022.

6 *"To see it in a single number was an epiphany"*: Video interview with Ken Rosenberg, Jan. 10, 2021.

7 *An estimated 50 million people consider themselves birdwatchers*: "2011 Survey of Fishing, Hunting and Wildlife-Associated Recreation National Overview," U.S. Fish and Wildlife Service National Digital Library website, Aug. 2012, https://digitalmedia.fws.gov/digital/collection/document/id/859. Note: The number of hunters of all game has been decreasing over the past decade, according to the U.S. Fish and Wildlife Service, but the sport also experienced a jump in interest during the pandemic that hasn't been measured. Estimates on the overall number of hunters—most of them deer hunters—range from 11 to 15 million. Ibid. As for the number of waterfowl hunters, there were 1,591,797 of the required federal Duck Stamps sold in 2020–2021. That number has slowly and steadily decreased from the 1975–1976 sales of 2,207,318. "Duck Stamps Sold Year by Year," U.S. Fish and Wildlife Service website, https://www.fws.gov/media/duck-stamp-sales-year.

8 *Our favorite headline came from the online magazine* Slate: Nicholas Lund, "You Have No Choice but to Become a Backyard Birder," *Slate*, March 28, 2020, https://slate.com/technology/2020/03/how-to-bird-during-pandemic.html.

8 *As the nation boomed, we harvested the bulk of our old-growth forests:* "Are Old-Growth Forests Protected in the U.S.?" *Scientific American*, Jan. 22, 2009, https://www.scientificamerican.com/article/are-old-growth-forests/.

8 *plowed up 60 percent of the continent's grasslands:* "The Plowprint Report: 2021," World Wildlife Fund, Sept. 14, 2021, https://www.worldwildlife.org/projects/plowprint-report.

8 *and drained more than half of our wetlands:* "Wetlands Losses in the United States, 1780s to 1980s. Report to Congress," OSTI.gov, U.S. Department of Energy. Jan. 1, 1990, https://www.osti.gov/biblio/5527872.

1: On the Edge of Extinction

12 *This unique subspecies of grasshopper sparrows:* "Recovery Plan for the Florida Grasshopper Sparrow," U.S. Fish and Wildlife Service website, May 18, 1999, https://ecos.fws.gov/docs/recovery_plan/Florida%20Grasshopper%20Sparrow%20Recovery%20Plan%20Amendment_1.pdf.

12 *Then in 2003, a routine survey:* Interviews with members of the Grasshopper Sparrow working group between 2019 and 2022.

12 *"The population just completely crashed":* Phone interview with Reed Bowman, Jan. 27, 2020.

12 *Finally so few sparrows remained—just twenty-two breeding pairs altogether:* Ibid.

13 *"We had zero experience":* Interview with Steve Shurter, Yulee, Fla., Feb. 2, 2021.

13 *Meanwhile, candidates are stacking up:* Michael Wines, "Endangered or Not, but at Least No Longer Waiting," *New York Times*, March 6, 2013, https://www.nytimes.com/2013/03/07/science/earth/long-delayed-rul ings-on-endangered-species-are-coming.html.

14 *The first was the late Howard Gilman:* "The Fall of the House of Gilman," *Forbes*, Aug. 11, 2003, https://www.forbes.com/forbes/2003/0811/068 .html?sh=46df59fc2086.

14 *Shortly after taking control of the family business in 1982:* Ibid.

14 *Isabella Rossellini, the Italian actress and once the highest-paid model:* Angelica Jade Bastién, "Isabella Rossellini Has Defined—and Redefined—Beauty for Decades. She Doesn't Intend to Stop Now," *Harper's BAZAAR*, May 3, 2021, https://www.harpersbazaar.com/culture/fea tures/a36182114/isabella-rossellini-interview/.

14 *Bill Clinton escaped to White Oak:* "The Fall of the House of Gilman," *Forbes*.

14 *Gilman was a close friend and patron of Baryshnikov:* Alan M. Kriegsman, "On the Road with Mark and Misha," *Washington Post*, July 28, 1991, https://www.washingtonpost.com/archive/lifestyle/style/1991/07/28 /on-the-road-with-mark-and-misha/0b7b4190-b3c2-4dbc-95ef-c7072 97c75a3/; Joan Acocella, "The Soloist. At fifty, Mikhail Baryshnikov Reflects on How Ballet Saved Him," *The New Yorker*, Jan. 11, 1998, https:// www.newyorker.com/magazine/1998/01/19/the-soloist.

15 *a favorite pastime was feeding Gilman's baby giraffes:* Kriegsman, "On the Road with Mark and Misha."

15 *He'd accumulated $550 million in debt:* "The Fall of the House of Gilman," *Forbes*.

15 *Eventually, White Oak went up for sale*: Robbie Whelan, "End of an Era for White Oak Plantation," *Wall Street Journal*, from a video on its "The News Hub" platform, April 6, 2012, https://www.wsj.com/video/end -of-an-era-for-white-oak-plantation/1EBE7FC5-FF30-428C-A8D7 -462BF6152870.html.

15 *In 2013, White Oak got its second billionaire*: "Dodgers Owner Buys White Oak Plantation," *Jacksonville Times-Union*, March 20, 2013.

16 *"We just couldn't find a female":* Phone interview with Michael Delaney, March 24, 2021.

16   *The last Dusky died in captivity at Disney World in 1987:* Cass Peterson, "Goodbye Dusky Seaside Sparrow," *Washington Post*, June 18, 1987, https://www.washingtonpost.com/archive/politics/1987/06/18/goodbye -dusky-seaside-sparrow/b8c8d618-54eb-4f17-a10c-3fbae9eb0672/.

16   *the Fish and Wildlife Service declared the species extinct:* "Dusky Seaside Sparrow," Endangered List, https://endangeredlist.org/animal/dusky-sea side-sparrow/.

16   *Rare Species is a very different place:* Interview with Paul Reillo, Loxahatchee, Fla., Feb. 19, 2021.

16   *"I reached out to pretty much every program":* Interview with Andrew Schumann, Yulee, Fla., Feb. 2, 2021.

17   *"They were perfect little birds":* Interview with Paul Reillo, Loxahatchee, Fla., Feb. 19, 2021.

17   *"We weren't really sure if they knew":* Interview with Juan Oteyza, Florida prairie near Kissimmee, Fla., Feb. 9, 2021.

17   *Ninety percent of the dry Florida prairie ecosystem is gone:* Meghan Bartels, "Meet the Author: Q & A with . . . Reed Noss!" *Island Press* website, May 21, 2015, https://islandpress.org/blog/meet-author-q-reed-noss.

18   *"He's on his own now":* Interview with Juan Oteyza, Florida prairie near Kissimmee, Fla., Feb. 9, 2021.

18   *"By the time the first wave of sparrows lands on the prairie":* Interview with members of Florida Grasshopper Sparrow working group in 2020 and 2021.

19   *"This is our bird":* Phone interview with Paul Gray, Jan. 27, 2020.

19   *"There's one there":* Interview with Sarah Biesemier, Florida prairie near Kissimmee, Fla., March 15, 2021.

20   *The flock on the prairie produced more than a hundred fledglings:* Interviews with Juan Oteyza, in person in Florida and by phone over the course of 2022.

20   *"The outlook is so much better":* Interview with Juan Oteyza, Feb. 10, 2021.

21   *On February 11, 2004, in northeast Arkansas:* This description of events is drawn from several sources. They include the *60 Minutes* archive titled "The Lord God Bird," 2005; a *Washington Post* report by David Brown and Eric Pianin titled "Extinct? After 60 Years, Woodpecker Begs to Differ," April 29, 2005; and Angelita Faller, "UA Little Rock Research Recalls Search for Once Extinct Ivory-billed Woodpecker That Is Now Back Up for Extinction," University of Arkansas website, Oct. 6, 2021, https://ualr.edu/news/2021/10/06/luneau-ivory-billed-woodpecker/.

21   *The Ivory-bill, as locals call it*: American Bird Conservancy website, June 21, 2019, https://abcbirds.org/bird/ivory-billed-woodpecker/.

21   *It also drums a tree in a telltale pattern*: Frank J. Severson, "Memories from the Singer Tract," *Birding Magazine*, March/April 2007.

22   *Even though the woodpeckers hadn't been captured in a clear photo*: Stephen Lyn Bales, "A Close Encounter with the Rarest Bird," *Smithsonian Magazine*, Sept. 2010, https://www.smithsonianmag.com/science-nature/a-close-encounter-with-the-rarest-bird-54437868/.

22   *and the last undisputed sighting occurred in April 1944*: "Recovery Plan for the Ivory-billed Woodpecker," U.S. Fish and Wildlife Service, April 16, 2010, https://www.nrc.gov/docs/ML1430/ML14309A100.pdf.

22   *The searchers weren't even allowed to tell their families"*: Scott Weidensaul, "Ghost of a Chance," *Smithsonian Magazine*, Aug. 2005, https://www.smithsonianmag.com/science-nature/ghost-of-a-chance-82491331/.

22   *On April 28, 2005, Interior Secretary Gail Norton*: Brown and Pianin, "Extinct? After 60 Years, Woodpecker Begs to Differ."

22   *"I can't begin to tell you"*: Ibid.

22   *"Amazingly, America may have another chance to protect"*: Cornell University, "Long Thought Extinct, Ivory-billed Woodpecker Rediscovered in Big Woods of Arkansas," *Science Daily*, April 28, 2005, https://www.sciencedaily.com/releases/2005/04/050428094235.htm.

23   *The media coverage was exhaustive*: Greg Allen, "Tourism Grows Around Ivory-billed Woodpecker," *Morning Edition*, NPR, Dec. 26, 2005.

23   *Penny Child's hair salon added a $25 "Ivory-bill Haircut"*: Ibid.

23   *Over the next four years the search*: "Recovery Plan for the Ivory-billed Woodpecker," U.S. Fish and Wildlife Service.

23   *Tips of sightings and suggestions poured in*: Ibid.; Kenneth Heard, "$50,000 Offered for Proof of Bird's Existence," *Arkansas Democrat-Gazette*, Nov. 9, 2008, https://www.arkansasonline.com/news/2008/nov/09/50000-offered-proof-birds-existence-20081109/.

23   *In the end, after some 20,000 collective hours*: "The Ivory-bill After a Decade: $20.3 Million Spent, Total Cost 'Unknown,'" BirdWatching website, Feb. 2015, https://www.birdwatchingdaily.com/news/conservation/the-ivory-bill-after-a-decade-20-3-million-spent-total-cost-unknown/.

23   *the best evidence of Ivory-billed Woodpeckers*: "Recovery Plan for the Ivory-billed Woodpecker," U.S. Fish and Wildlife Service.

23   *a 156-page recovery plan*: "Recovery Plan for the Ivory-billed Woodpecker," U.S. Fish and Wildlife Service, April 16, 2010, https://www.nrc.gov/docs/ML1430/ML14309A100.pdf.

23 *But the most significant impact turned out to be*: Jerome A. Jackson, "Ghost Bird: A Look Back at Ivory-bill Fever," BirdWatching website, Feb. 1, 2015, https://www.birdwatchingdaily.com/news/species-profiles/ghost -bird-a-look-back-at-ivory-bill-fever/.

24 *"I'm going to tell you something that has not been published"*: Interview with John Fitzpatrick, Vero Beach, Fla., Feb, 18, 2022.

25 *"It's real. It's out there."*: Interview with Steve Latta, Central Louisiana, April 11, 2021.

25 *"They got wind of it"*: Video interview with Mark Michaels, March 4, 2021.

26 *This is where the Ivory-bills of long ago liked to nest*: James T. Tanner, *The Ivory-Billed Woodpecker* (Mineola, N.Y.: Dover, 1942), 17.

27 *"I don't know where it came from"*: Interview with Steve Latta, Central Louisiana, April 12, 2021.

27 *"It flew at an angle so we had a much longer look"*: Interview with Tommy Michot, Central Louisiana, April 12, 2021.

28 *"I don't typically expect people to believe me"*: Interview with Peggy Shrum, Central Louisiana, April 12, 2021.

28 *"There's a wild boar"*: Interview with Tommy Michot, Central Louisiana, April 12, 2021.

29 *"That's turned out to be more difficult"*: Interview with Steve Latta, Central Louisiana, April 12, 2021.

29 *Since the last confirmed view in 1944:* B. J. Hollars, "Unsolved Histories: A Bird Lost, a Sketch Found, and a Dream to Bring It Back," *Michigan Quarterly Review*, January 2016.

29 *In all that time, not one unequivocal photo:* Matthew Brown, "U.S. Says Ivory-billed Woodpecker, 22 Other Species Extinct," Associated Press, Sept. 29, 2021.

29 *"That was probably one of the hardest things I've done"*: Catrin Einhorn, "Protected Too Late: U.S. Officials Report More than 20 Extinctions," *New York Times*, Sept. 28, 2021.

30 *"The hearing was a joke"*: Interview with John Fitzpatrick, Feb. 18, 2022.

30 *The most detailed response to the service's proposal:* Steve C. Latta et al., "Multiple Lines of Evidence Indicate Survival of the Ivory-billed Wood-pecker in Louisiana," bioRxiv.org, April 6, 2022.

31 *Dave Naugle, a wildlife biology professor at the University of Montana:* Video interview with Dave Naugle, July 1, 2021.

31 *At the start of 2023, there were about a hundred birds:* "FWS-Listed U.S. Species by Taxonomic Group—All Animals," Environmental Conserva-

tion Online System, U.S. Fish and Wildlife Service website, https://ecos
.fws.gov/ecp/report/species-listings-by-tax-group?statusCategory=Listed
&groupName=Birds&total=108. Note: On September 30, 2021, the U.S.
Fish and Wildlife Service proposed delisting twenty-three species, including eleven birds—eight of them from Hawaii—from the Endangered
Species List due to extinction. As of this printing, the final rulings had not
been announced.

31   *"The way that public money has been spent has been to put out fires":* Interview with Scott Sillett, Takoma Park, Md., Oct. 12, 2021.

2: Vanishing by the Billions

33   *For weeks, Adam Smith had been crunching:* Video interview with Adam
Smith, March 31, 2021.

33   *"Well, that can't be right":* Ibid.

34   *"At any given time":* Corey T. Callagan, Nakagawa Shinichi, and William
K. Cornwell, edited by Simon Asher Levin, "Global Abundance Estimates for 9,700 Bird Species," Proceedings of the National Academy of
Science website, March 28, 2021, https://doi.org/10.1073/pnas.20231
70118.

34   *That's the number of breeding birds in billions:* Kenneth V. Rosenberg et al.,
"Decline of the North American Avifauna," *Science*, Sept. 19, 2019.

35   *"It always takes a couple of times":* Video interview with Adam Smith,
March 31, 2021.

35   *The 80,000-square-foot research center is a global mecca:* "Cornell Ornithology Laboratory / RMJM," *ArchDaily*, April 16, 2009, https://www.arch
daily.com/19263/cornell-ornithology-laboratory-rmjm.

35   *Whether you're entering the building for a tour:* Details on the Cornell Lab,
its building, and materials were provided in a series of interviews with
Cornell Lab staff writer Pat Leonard, two visits to the lab in 2019 and
2021, and the lab's website.

36   *During that July conference, the prestigious core group:* Partners in Flight
Science Committee meeting notes, July 11–13, 2017.

36   *because more than forty varieties live in North America:* Marc Devokaitis,
"Learn How to ID These 5 Confusing Streaked Sparrows," All About
Birds website, Oct. 8, 2019, https://www.allaboutbirds.org/news/learn
-how-to-id-these-5-confusing-streaked-sparrows/#.

37   *"Exactly how many birds have we lost altogether":* Interviews with Ken
Rosenberg, John Sauer, Peter Blanchard, and Mike Parr, conducted from
January 2021 through September 2021.

37 *"Mike was really pushing us":* Video interview with Ken Rosenberg, March 17, 2021.

37 *Over millions of years:* The total number of species in the world is estimated at nearly 11,000, but it's a moving target because the various international lists count birds in different ways. Some species are added and subtracted based on biological and genomic research that has grown increasingly sophisticated in assessing when a bird is sufficiently distinct to make up its own species. The 11,000 number comes from BirdLife International, which keeps the official "Red List" that documents bird families and extinctions. "State of the World's Birds 2021 Update," BirdLife International, 2021, http://datazone.birdlife.org/2021-annual-update.

37 *Every fall and spring, the migrators:* "The State of North America's Birds 2016," The North American Bird Conservation Initiative, 2016, https://www.stateofthebirds.org/2016/.

38 *Contrary to the bird-brained label:* Jennifer Ackerman, *The Genius of Birds* (New York: Penguin, 2016); Jennifer Ackerman, *The Bird Way: A New Look at How Birds Talk, Work, Play, Parent and Think* (New York: Penguin, 2020).

38 *Their flight skills are astounding:* Scott Weidensaul, *A World on the Wing: The Global Odyssey of Migratory Birds* (New York: W. W. Norton, 2021).

38 *Even the love lives of birds:* Richard O. Prum, *The Evolution of Beauty: How Darwin's Forgotten Theory of Mate Choice Shapes the Animal World—and Us* (New York: Doubleday, 2017).

38 *Depending on the species, as many as 90 percent:* Phone interview with John Fitzpatrick, July 18, 2022.

38 *Even so, by the fall when a new generation takes flight:* Interview with Pete Marra and Ken Rosenberg, Ithaca, N.Y., Aug. 31, 2021.

38 *an equilibrium, thought for the past fifty years to be roughly 10 billion:* Ibid.

38 *Throughout tens of thousands of years, even millions in some cases:* Irby J. Lovette and John W. Fitzpatrick, *Handbook of Bird Biology* (Princeton: Princeton University Press, 2004), Chapter Three.

38 *The five-inch-long Golden-cheeked Warbler:* Interview with U.S. Army biologist John Macey, Killeen, Tex., May 3, 2021.

39 *A major study by the National Audubon Society:* "Survival by Degrees: 389 Bird Species on the Brink," National Audubon Society website, https://www.audubon.org/climate/survivalbydegrees.

39 *Collisions with high-rise glass-covered buildings:* Priyanka Runwal, "Building

Collisions Are a Greater Danger for Some Birds than Others," National Audubon Society website, July 9, 2020, https://www.audubon.org /news/building-collisions-are-greater-danger-some-birds-others.

39 *Outdoor cats kill an average of 2.6 billion birds:* "Threats to Birds," U.S. Fish and Wildlife Service website, https://www.fws.gov/library/collections /threats-birds. Estimates for mortality from cats range from 1.4 to 3.9 billion. The most detailed accounting can be found in Peter P. Marra and Chris Santella, *Cat Wars: The Devastating Consequences of a Cuddly Killer* (Princeton: Princeton University Press, 2016).

39 *In still preliminary estimates, wind turbines are thought to kill:* Joel Merriman, American Bird Conservancy website, Jan. 26, 2021, https://abcbirds .org/blog21/wind-turbine-mortality/?gclid=Cj0KCQjw39uYBhCLAR IsAD_SzMQALyuIOLLW2i-rxeOmpClkGW-kV_REfvhcq95o0KrjB T3UxQ27glIaAu_IEALw_wcB.

39 *Pesticides, and in particular a newer class of insecticides called neonicotinoids:* "EPA Confirms Three Widely Used Neonicotinoid Pesticides Likely Harm Vast Majority of Endangered Plants, Animals," Center for Biological Diversity, June 16, 2022, https://biologicaldiversity.org/w/news/press -releases/epa-confirms-three-widely-used-neonicotinoid-pesticides -likely-harm-vast-majority-of-endangered-plants-animals-2022-06-16/.

39 *But the Environmental Protection Agency, which is responsible:* "Balancing Wildlife Protection and Responsible Pesticide Use: How EPA's Pesticide Program Will Meet Its Endangered Species Obligations," Environmental Protection Agency website, April 2022, https://www.epa.gov/system /files/documents/2022-04/balancing-wildlife-protection-and-responsi ble-pesticide-use_final.pdf.

39 *although there's little question they are a contributing factor:* Scott Weidensaul, "Neonic Nation," *LivingBird Magazine*, Summer 2022.

39 *These were the issues on Mike Parr's mind:* Interviews with Mike Parr over the course of 2021 and 2022.

40 *Since 2017 when Parr took the helm as ABC's second president:* Interview with Mike Parr, Jan. 22, 2022.

41 *"A number of us in that room were skeptical":* Interview with Tom Will, Washington, D.C., Oct. 10, 2021.

41 *"I remember somebody said":* Interview with Mike Parr, Washington, D.C., Oct. 13, 2021.

41 *"It's such a simple question":* Interview with Adriaan Dokter, Ithaca, N.Y., Aug. 30, 2021.

42 *Then they used a secret sauce:* Ibid.

42 *"It matched perfectly":* Ibid.

42 *In the spring of 2019, Adam Smith:* Video interview with Adam Smith, Oct. 21, 2021.

42 *"It was a eureka moment":* Ibid.

42 *Dozens of back-and-forth emails:* The emails attributed here as well as the chronology of this research project were reconstructed from interviews, emails, and Partners in Flight documents collected over 2021 and 2022 from the key participants.

43 *"We've got to get this into* Science": Interview with Ken Rosenberg, Ithaca, N.Y., Aug. 30, 2021.

44 *"It's important to get this published where it has credibility":* Interview with Pete Marra, Washington, D.C., Oct. 12, 2021.

44 *Marra is a workhorse:* Interviews with Pete Marra, Ken Rosenberg, and other authors of the report during 2021.

44 *"Slowing the loss of biodiversity is one of the defining environmental challenges":* Rosenberg et al., "Decline of the North American Avifauna," *Science,* Oct. 4, 2019.

45 *"I started to cry":* Video interview with Miyoko Chu, June, 21, 2021. Additional material in this section came from a follow-up interview on Sept. 1, 2021, in Ithaca, N.Y.

46 *They built a website:* "3 Billion Birds Gone," https://www.3billionbirds.org.

46 *The headlines were bold and succinct:* Carl Zimmer, "Birds Are Vanishing from North America," *New York Times,* Sept. 19, 2019.

46 *NBC News reported:* Jeremy Deaton, "U.S., Canada Have Lost 3 Billion Birds Since 1970," NBC News website, Sept. 19, 2019.

46 *Mike Parr was in Monterey, California:* Video interview with Mike Parr, Jan. 9, 2021.

47 *A shy, gentle introvert with thick black glasses:* Linda Leer, *Rachel Carson: Witness for Nature* (New York: Houghton Mifflin Harcourt, 1997), 227.

47 *"How could intelligent beings seek to control":* Rachel Carson, *Silent Spring* (New York: Houghton Mifflin, 1962), 56.

48 *"I do think this could be the next* Silent Spring": Interview with Pete Marra, Ithaca, N.Y., Aug. 30, 2021.

48 *"Silent Spring changed the playing field":* Interview with Ken Rosenberg, Ithaca, N.Y., Aug. 30, 2021.

48 *The Fish and Wildlife Service, headquartered:* "The U.S. Fish and Wildlife

Service," Congressional Research Service, July 20, 2018, https://www
.everycrsreport.com/reports/R45265.html.

48 *Audubon has the largest membership:* Video interview with Audubon's Elizabeth Gray, Sept. 16, 2021.

48 *with annual revenues over $340 million:* Video interview with Ducks Unlimited's Adam Putnam, Aug. 23, 2021.

49 *"Yeah, that's about right":* Interview with Ian Owens, Ithaca, N.Y., Aug. 30, 2021.

49 *"I would say we have a decade to get this right":* Video interview with Elizabeth Gray, Sept. 16, 2021.

3: Era of Discovery

51 *our national symbol, adopted by the Second Continental Congress*: "Our National Symbol," American Eagle Foundation website, https://www.eagles .org/what-we-do/educate/.

52 *In the mid-1900s, the widespread use of pesticides*: Jack E. Davis, "The Bald Eagle's Soaring Return Shows That the U.S. Can Change for the Better," *Smithsonian Magazine*, April 2022. Note: Some assessments had the lowest eagle population at 417 in the lower forty-eight states in 1963, https://www.smithsonianmag.com/science-nature/bald-eagle-soaring -return-shows-us-can-change-180979798/.

52 *The eagle began a slow recovery:* "History of the Bald Eagle Decline, Protection and Recovery," U.S. Fish and Wildlife Service.

52 *The Bald Eagle was one of the first birds protected*: Lynda V. Mapes, "Bald Eagle's Recovery Is a Triumph—for Now," *Seattle Times* website, June 28, 2007, https://www.seattletimes.com/seattle-news/bald-eagles -comeback-is-a-triumph-8212-for-now/.

52 *A celebration was held at the Jefferson Memorial:* Meghan Vittrup, "Defense Department Helps Eagle Soar off Endangered List," American Eagle Foundation website, https://www.eagles.org/defense-department-helps -eagle-soar-off-endangered-list/.

52 *When it came time to compile the latest eagle count*: "New Bald Eagle Population Estimate Incorporates eBird data," eBird.org website, March 24, 2021, https://ebird.org/news/new-bald-eagle-population-estimate-usfws.

53 *The final tally, announced in 2021*: Derrick Bryson Taylor, "America's Bald Eagle Population Has Quadrupled," *New York Times*, March 25, 2021. Note: The Bald Eagle's recovery has come despite the spread of lead poisoning in more than half of the Bald and Golden Eagles tested in a comprehensive study released in 2022. The Golden Eagle, whose population

isn't growing as fast as the Bald Eagle, is more threatened by the poisoning. But the lead, which is ingested by the birds from carrion left in the field by hunters using lead bullets, can be lethal for both species. Pressure has mounted to ban the use of lead bullets, which are prohibited in duck hunting, and require a switch to copper or steel bullets. Outgoing U.S. Fish and Wildlife director Dan Ashe issued an order to phase out lead bullets in 2016, but the edict was immediately reversed by incoming interior secretary Ryan Zinke with the Trump administration. The argument against banning lead is that hunters are more likely to switch to other ammunition if it's a voluntary move. U.S. Geological Survey biologist Todd Katzner, who helped oversee the study on lead poisoning in eagles, said no one who witnesses the impact lead has on these birds would want to hunt with lead bullets again. "It's one of the most unpleasant things you'll ever see," he said. "We think of the Bald Eagle as majestic. It's on our money and all that. But here's this bird, its feet all curled up. It can't even stand."

53 *A lab in Colorado Springs collects DNA from feathers throughout the hemisphere:* Video interview with Kristen Ruegg, head of the Bird Geoscape Project at Colorado State University, on Oct. 19, 2021.

53 *In Hawai'i, miniature monitors that work like supermarket checkout scanners:* Interview with André Raine, Hanapepe, HI, June 4, 2021.

54 *The year was 1870, a time when:* Malcolm Smith, "A Hatful of Horror: The Victorian Headwear Craze That Led to Mass Slaughter," *History Extra, BBC History Magazine* website, Feb. 12, 2021, https://www.historyextra.com/period/victorian/victorian-hats-birds-feathered-hat-fashion/.

54 *a fashion craze for elaborate plumes on women's hats:* "The Victorian Penchant for Plumage," 19thcenturyghosts.com, https://19thcenturyghosts.com/2015/12/15/the-victorian-penchant-for-plumage/.

54 *sold by the ounce, were double the price of gold:* Laura Allen, "The Price of a Feather," National Parks Conservation Association website, Summer 2015, https://www.npca.org/articles/918-the-price-of-a-feather.

54 *As a teenager, Bradley got a job as a guide for plume hunters in South Florida:* Maureen Sullivan-Hartung, "Hats off to Guy Bradley and Other Everglades Wardens," *Naples Florida Weekly,* June 27, 2013.

54 *Hunters often shot them on their nests:* John Dolen, "Guy Bradley Was a Poacher-Turned-Gamekeeper Who Gave His Life in Defense of Some of South Florida's Most Colorful Creatures," *Fort Lauderdale Magazine* website, July 1, 2016, https://fortlauderdalemagazine.com/plume-bird-martyr/.

54 *When Congress cracked down with the Lacey Act:* Ibid.; Frank Graham Jr.,

The Audubon Ark: A History of the National Audubon Society (New York: Alfred A. Knopf, 1990), 23.

54 *The next day, a search party:* Ibid., 58.

55 *Even though the evidence showed Bradley's gun remained fully loaded:* Ibid.; Jack E. Davis, *An Everglades Providence: Marjory Stoneman Douglas and the American Environmental Century* (Athens: University of Georgia Press, 2009), 191.

55 *Meanwhile, as the turn of the century approached:* Linton Weeks, "Hats Off to Women Who Saved the Birds," NPR History Department website, July 15, 2015, https://www.npr.org/sections/npr-history-dept/2015/07/15/422860307/hats-off-to-women-who-saved-the-birds.

55 *The Migratory Bird Treaty Act of 1918:* Ibid.

55 *Unregulated hunting for food and the drought of the Dust Bowl years:* Matthew L. Miller, "Epic Duck: The Story of the Canvasback," Cool Green Science blog of The Nature Conservancy, Oct. 16, 2017, https://blog.nature.org/science/2017/10/16/epic-duck-story-canvasback-birds/.

55 *Their influence helped push through a series of actions in Congress:* Lisa Irby, "Celebrating 80 Years of the Pittman-Robertson Act," Ducks Unlimited website, https://www.ducks.org/conservation/public-policy/celebrating-80-years-of-the-pittman-robertson-act.

55 *In the years since, financial support for waterfowl:* "Firearm Industry Surpasses $15 Billion in Pittman-Robertson Excise Tax Contributions for Conservation," Firearm Industry Trade Association, May 5, 2022, https://www.nssf.org/articles/firearm-industry-surpasses-15-billion-in-pittman-robertson-excise-tax-contributions-for-conservation/.

55 *for the country's 567 National Wildlife Refuges:* "Visit a National Wildlife Refuge Facility," U.S. Fish and Wildlife Service website, https://www.fws.gov/visit-us.

56 *In the early 1960s, a birdwatcher from the Midwest:* R. Michael Erwin and Robert Blohm, "Migratory Bird Program at the U.S. Geological Survey Patuxent Wildlife Research Center/U.S. Fish and Wildlife Service Patuxent Research Refuge: Transformations in Management and Research," Publications of the U.S. Geological Survey, 2016, https://digitalcommons.unl.edu/usgspubs/142/.

56 *Robbins came up with an ambitious plan to build a databank:* Documents compiled for authors by the U.S. Geological Survey on annual bird banding from 1962 through 2021.

56 *Danny Bystrak, a lifelong biologist at Patuxent:* Interview with Danny Bystrak, Laurel, Md., Oct. 15, 2021.

56 *"You're doing like fifty stops":* Ibid.

57 *"This guy was the dynamo of twentieth-century ornithology":* Interview with John Sauer, Laurel, Md., Oct. 15, 2021.

57 *When the Department of the Interior called for a comprehensive report:* "North American Breeding Bird Survey," Eastern Ecological Science Center, U.S. Geological Survey website, March 19, 2018, https://www.usgs.gov/cen ters/eesc/science/north-american-breeding-bird-survey.

57 *"Suddenly you saw the real power of fifty years of data":* Interview with Danny Bystrak, Laurel, Md., Oct. 15, 2021.

57 *Smith, the scientist who did the report's final calculations:* Video interview with Adam Smith, Feb. 10, 2021.

57 *An electrical engineer from Indiana named Bill Cochran:* Jessie Greenspan, "Chasing Birds Across the Country—for Science," National Audubon Society website, Oct. 15, 2015, https://www.audubon.org/news/chasing -birds-across-countryfor-science.

58 *"We are dealing with a different beast now":* Video interview with Pete Marra, Feb. 12, 2021.

58 *A photo from forty years ago shows John Fitzpatrick:* Scott Weidensaul, "The 'Remarkably and Persistently Stimulating' John Fitzpatrick," *LivingBird Magazine*, April 2, 2021.

58 *"Binoculars, a notebook, all done by hand":* Interview with John Fitzpatrick, Ithaca, N.Y., Aug. 31, 2021.

58 *Fitz, as everyone calls him, took the director post:* Interview with John Fitzgerald, Vero Beach, Fla., Feb. 18, 2022.

59 *He grew the lab of about forty scientists:* Several interviews with John Fitz-patrick over 2021 and 2022.

60 *Every year since the first eagles built a nest:* Dana Hedgpeth, "Baby Eaglet Has Died, Other Egg Starts to Hatch at National Arboretum," *Washington Post*, March 28, 2022, https://www.washingtonpost.com/dc-md -va/2022/03/28/dead-eaglet-national-arboretum-dc/.

61 *Eagle cams peer into nests all over the country:* Mike Kilen, "Bald Eagles Go from Endangered Species to Reality Stars," *Des Moines Register*, April 28, 2014.

61 *providing millions of people with mesmerizing footage:* Jack E. Davis, *The Bald Eagle: The Improbable Journey of America's Bird* (New York: Liveright, 2022), 713.

61 *"I can write letters and tell people how cool birds are":* Kilen, "Bald Eagles Go from Endangered Species to Reality Stars."

61 *As Eldermire pointed out:* Ibid.

4: Following the Birds

64 *"That's everyone":* Interview with Reed Bowman, Lake Placid, Fla., Feb. 17, 2021.

64 *"We can't do anything if we don't know":* Video interview with Nathan Cooper, Nov. 17, 2021.

64 *The work is headquartered at Archbold:* "History—Beginnings: Founded in 1941 by Richard Archbold," Archbold Biological Station website, https://www.archbold-station.org/html/aboutus/history.html.

65 *The number of scrub-jays has gone from an estimated historical population of 40,000:* "Florida Scrub-Jays: 2019 by the Numbers," Audubon Florida website, https://fl.audubon.org/news/florida-scrub-jays-2019-numbers.

65 *to about 4,000:* Raoul K. Boughton and Reed Bowman, "Statewide Assessment of Florida Scrub-Jays on Managed Areas: A Comparison of Current Populations to the Results of the 1992–93 Survey," report submitted to the U.S. Fish and Wildlife Service, May 9, 2011. Provided to the authors by Reed Bowman.

66 *"This was a huge part of the puzzle":* Phone interview with Reed Bowman, March 2, 2020.

66 *"I said, 'Oh my God, look at all the data'":* Phone interview with Young Ha Suh, April, 2, 2020.

66 *"This is really cutting edge":* Phone interview with Reed Bowman, March 2, 2020.

67 *The Florida scrub is dotted with some of North America's oldest plants:* "Native Plant Communities, Scrub," Florida Native Plant Society website, https://www.fnps.org/natives/native-plant-community/scrub.

67 *with captivating names like Florida scrub rockrose:* Craig N. Huegel, PhD, "Florida Scrub Rockrose," Native Florida Wildflowers website, April 30, 2021, http://hawthornhillwildflowers.blogspot.com/2021/04/florida-scrub-rockrose-crocathemum.html.

67 *and nodding pinweed:* David Sedore, "Nodding Pinweed," Wild South Florida website, http://www.wildsouthflorida.com/nodding.pinweed.html.

67 *The most obvious plants are:* "Native Plant Communities, Scrub," Florida Native Plant Society website.

67 *On average, each pair of scrub-jays needs twenty-five acres:* Series of author interviews with Reed Bowman over 2020 and 2021.

68 *Hundreds of supporters:* "Jay Watch: Dedicated to Protecting the Florida Scrub-Jay, Our State's Only Endemic Native Bird," Audubon Florida website, https://fl.audubon.org/get-involved/jay-watch.

68 *Bowman got a taste of the ambivalence:* Interview with Reed Bowman, Lake Placid, Fla., Feb. 17, 2021.

68 *Fitzpatrick tells the story:* Interview with John Fitzpatrick, Lake Placid, Fla., Feb. 16, 2021.

70 *Passage of the Florida Wildlife Corridor Act:* "Conservation Florida Celebrates Signing of Florida Wildlife Corridor Act," Conservation Florida website, June 30, 2021, https://www.conserveflorida.org/cfl news/2021/6/30/conservation-florida-celebrates-signing-of-the-florida -wildlife-corridor-act.

70 *"The files are so big":* Interview with Young Ha Suh, Ithaca, N.Y., Aug. 31, 2021.

70 *"We found they're a lot like us":* Phone interview with Reed Bowman, March 2, 2020.

71 *"We're studying how to mimic nature":* Interview with John Fitzpatrick, Lake Placid, Fla., Feb. 16, 2021.

72 *"I absolutely recognize that there are places we're going to lose":* Interview with Reed Bowman, Lake Placid, Fla., Feb. 16, 2021.

72 *Banding is the earliest and simplest tracking form:* Bob Montgomerie, "Audubon's Legendary Experiments," American Ornithology Society website, June 13, 2018, https://americanornithology.org/audubons-leg endary-experiments/.

72 *In the spring of 1804:* Ibid.

72 *"We're in a renaissance of ornithology":* Alisa Opar, "Have Wings Will Travel," *Audubon Magazine,* Spring 2022, https://www.audubon.org /magazine/spring-2022/this-pioneering-collaboration-will-open-new.

73 *"We have to move fast so we don't traumatize them":* Interview with Antonio Celis-Murillo, Laurel, Md., Oct. 17, 2021.

73 *Now more than 6,000 licensed specialists:* Ibid.

75 *Zimorski, a biologist with the Louisiana Department of Wildlife and Fisher-ies:* Interview with Sara Zimorski, Gueydan, La., March 8, 2021.

75 *conversion of wetlands and grasslands to agricultural fields*: Deborah Fuller, "Whoopers Return to Louisiana After 60 Years," U.S. Fish and Wildlife Service, Endangered Species Program section, Jan. 18, 2012.

75 *Only twenty-one Whooping Cranes were living*: Ibid.; "Wisconsin Whoop-ing Crane Management Plan," Wisconsin Department of Natural Re-sources, Dec. 6, 2006.

75 *By January 2022, the program reached a milestone of eight hundred birds*: Phone interview with Sara Zimorski, June 16, 2022.

76 *Dozens of young cranes made the journey*: Lee Bergquist, "Federal Authorities

to End Use of Ultralights for Whooping Crane Project," *Milwaukee Journal Sentinel*, Jan. 23, 2016, https://archive.jsonline.com/news/wisconsin/fish-and-wildlife-service-to-end-whooping-crane-migration-project-b99657191z1-366304761.html/.

76 *As soon as they're old enough:* Interview with Sara Zimorski, Gueydan, La., March 8, 2021.

78 *Overall, the results in Louisiana are encouraging:* Ibid.

78 *Since the Louisiana project started in 2011, twelve birds have been shot:* Sara Sneath, "Poachers Threaten Louisiana's Mostly Successful Whooping Crane Reintroduction Program," NOLA.com, Sept. 6, 2020, https://www.nola.com/news/courts/article_2354dde4-eef4-11ea-aff0-eb82f52359ab.html.

78 *In one notorious case in 2016:* Associated Press, "Judge Orders Man to Pay $85K in Deaths of 2 Whooping Cranes," ABC News, July 31, 2020, https://abcnews.go.com/US/wireStory/judge-orders-man-pay-85k-deaths-whooping-cranes-72110494;"Federal Court Sentences Louisiana Man for Killing Whooping Cranes," *VL Outdoor Media*, July 31, 2020, http://vloutdoormedia.com/federal-court-sentences-louisiana-man-for-killing-whooping-cranes-you-better-read-this/.

79 *"Whooping Cranes can be the flagship species":* Interview with Sara Zimorski, Gueydan, La., March 8, 2021.

80 *The idea for the latest evolution of wildlife tracking:* Video interview with Martin Wikelski, Feb. 2, 2022.

80 *"You ecologists are stupid":* Sonia Shah, "Animal Planet: An Ambitious New System Will Track Scores of Species from Space," *New York Times Magazine*, Jan. 12, 2021.

80 *Not long afterward, Wikelski began":* Video interview with Martin Wikelski, Feb. 2, 2022.

80 *Motus is steadily expanding its towers:* Video interview with Stuart Mackenzie, Nov. 1, 2021.

81 *"People are really, actively following these animals":* Video interview with Martin Wikelski, Feb. 2, 2022.

82 *"This may be one of the best ways to understand":* Video interview with Stuart Mackenzie, Nov. 1, 2021.

82 *"There's a new technology every day":* Interview with Antonio Celis-Murillo, Laurel, Md., Oct. 15, 2021.

5: Listening to the Birds

85 *In 1917, when German U-boats were wreaking havoc:* Roy R. Manstan, *Cold Warriors, The Navy's Engineering and Diving Support Unit* (Bloomington, Ill.: AuthorHouse, 2014), Chapter Two; "History of Underwater Acoustics, World War I: 1914–1918," Discovery of Sound in the Sea website, https://dosits.org/people-and-sound/history-of-underwater-acoustics/world-war-i-1914-1918/.

85 *which included renowned inventor Thomas Edison:* E. David Chronon, "Thomas Edison: Unorthodox Submarine Hunter," Madison Literary Club and World War I website, Sept. 9, 1986, http://www.worldwar1.com/sfedsub.htm.

86 *"It was really primitive":* Phone interview with Dan Saenz, Feb. 7, 2020.

86 *Once Google and Facebook shared the software they created:* Series of interviews with Holger Klinck from August 2019 through September 2022.

86 *On the Big Island of Hawai'i:* Interview with Holger Klinck, Ithaca, N.Y., Sept. 1, 2022.

86 *At the Powdermill Nature Reserve near Pittsburgh:* Phone interview with Powdermill avian research coordinator Luke DeGroote, Oct. 29, 2019.

86 *In Africa, wildlife managers at certain national parks:* "Dying to Be Heard: Africa's Forest Elements Are Targets of Larger-Scale Acoustic Monitoring Effort," *Science Daily*, Oct. 14, 1999.

87 *"I'm well aware that you can talk to any scientist":* Phone interview with Justin Kitzes, Oct. 30, 2019.

87 *Connor Wood loaded his black Chevy pickup:* Video interview with Connor Wood, July 19, 2022.

89 *"Statistically gifted":* Video interview with Sarah Sawyer, May 20, 2021.

89 *"I went around and hooted at it from different distances":* Video interview with Connor Wood, April 16, 2021.

89 *The Sierra Nevada is a blend of breathtaking beauty:* "Our Region," Sierra Nevada Conservancy, State of California website, July 2020, https://sierranevada.ca.gov/about-us/.

89 *The range holds three national parks:* Ibid.

90 *Because of their preference for old-growth trees, owls are a barometer:* Interviews with Connor Wood and Sarah Sawyer by video and in person over the course of 2021 and 2022.

90 *until just several thousand of the birds are estimated to remain:* W. David Shuford and Thomas Gardali, eds., *California Bird Species of Special Concern* (2008), Chapter Two. Note: The precise number of California Spotted Owls is not known. The most detailed recent accounting published in

2008 said about 3,050 California owls, 4,779 Northern owls, and 1,592 Mexican owls remained. The California subspecies is thought to be losing about 2 percent of its population a year, which would put the current totals between 2,000 and 3,000 individuals.

90 *It didn't take long to piece together a picture*: Interviews by video and in person with Connor Wood, Ithaca, N.Y., over 2021 and 2022.

91 *Spotted Owls mate for life:* "Spotted Owl Life History," All About Birds website, https://www.allaboutbirds.org/guide/Spotted_Owl/lifehistory.

92 *"We thought, if we're going to do something":* Interview with Sarah Sawyer, Sacramento, Calif., May 20, 2021.

92 *"It's a balancing act":* Ibid.

93 *Three subspecies of Spotted Owl*: Ho Yi Wan et al., "Managing Emerging Threats to Spotted Owls," *Journal of Wildlife Management,* May 2018, https://www.researchgate.net/figure/Ranges-of-the-3-spotted-owl -subspecies-in-North-America-NSO-14-northern-spotted-owl-CSO _fig1_322868230.

93 *"As the traditional lumberjacks died off":* "Paul Bunyan, America's Best-Known Folk Hero," Wisconsin Historical Society website, https://www .wisconsinhistory.org/Records/Article/CS504.

93 *At the height of the forest harvesting:* Ted Gup, "Owl vs. Man: Who Gives a Hoot," *Time,* June 25, 1990, https://content.time.com/time/subscriber /article/0,33009,970447,00.html.

93 *A young biologist with the U.S. Fish and Wildlife Service:* "Eric Forsman Oral History interview," Oregon State University, Dec. 5, 2016, http://scarc.library.oregonstate.edu/omeka/exhibits/show/forestryvoices /item/34846.

94 *It took almost a decade, and years of legal battles:* Victor M. Sher, "The Spotted Owl's Journey Through the Federal Courts," *Public Land and Resources Law Review,* June 1993.

94 *The reaction was explosive*: Aaron Scott, "Timber Wars," Oregon Public Radio, October 2020.

94 *The owl made the cover of* Time *magazine*: Gup, "Owl vs. Man: Who Gives a Hoot."

94 *Death threats came in against Smokey Bear and Woodsy Owl:* Steven Lewis Yaffee, *The Wisdom of the Spotted Owl: Policy Lessons for a New Century* (Washington, D.C.: Island Press, 1994), xv.

94 *In 1993, President Bill Clinton, Vice President Al Gore:* Ibid., 141–43.

95 *President Donald Trump's administration lifted prohibitions:* Gillian Flaccus, "Trump Administration Slashes Imperiled Spotted Owls'

Habitat," Associated Press, Jan. 14, 2021, https://apnews.com/article/donald-trump-wildlife-washington-oregon-environment-7b9c53a88608054d521ac7d6025461fa.

95   *a ruling reversed by President Joe Biden the following year:* Gabrielle Canon, "Victory for the Spotted Owl as Trump-era Plan to Reduce Habitat Is Struck Down," *The Guardian*, Nov. 9, 2021.

95   *In Canada, the Spotted Owl has all but disappeared:* Cara McKenna, "How Canada Is Trying to Protect Its Last Three Spotted Owls," *The Guardian*, April 16, 2021, https://www.theguardian.com/environment/2021/apr/16/canada-last-three-spotted-owls.

95   *"It's not looking very good":* Warren Cornwall, "As Spotted Owls Numbers Keep Falling, Some Fear It's Doomed," *Seattle Times*, Aug. 13, 2008, https://www.seattletimes.com/seattle-news/as-spotted-owls-numbers-keep-falling-some-fear-its-doomed/.

97   *"It's safe to say":* Interview with Sarah Sawyer, Yosemite National Park, Calif., July 14, 2021.

97   *"That was really, really good news":* Interview with Connor Wood, Ithaca, N.Y., Sept. 1, 2021.

97   *Kelly tells the story:* Interview with Kevin Kelly, Stanislaus National Forest, Calif., July 14, 2021.

98   *"No one wants to be shooting owls":* Interview with Connor Wood, Ithaca, N.Y., Sept. 1, 2021.

98   *Nothing is left to chance:* Video interview with Nick Kryshak and Danny Hofstadter, Oct. 10, 2021.

99   *"We manage species using lethal methods all the time":* Video interview with Zach Perry, Nov. 3, 2021.

100   *"If we don't do these removals, the Spotted Owl will go extinct":* Video interview with Danny Hofstadter, Oct. 10, 2021.

100   *It also helps that they caught the invasion early:* Video interview with Connor Wood, July 19, 2022.

100   *After three years of shooting:* Connor Wood et al., "Early Detection of Rapid Barred Owl Population Growth Within the Range of the California Spotted Owl Advises the Precautionary Principle," Oxford Academic, Feb. 2020, https://academic.oup.com/condor/article/122/1/duz058/5670813.

100   *"This is such a special place for so many people":* Video interview with Connor Wood, July 19, 2022.

102   *She stands up, takes a deep breath:* Interview with Hilary Swain, Lake Placid, Fla., Feb. 17, 2021.

102 *In Yosemite Sarah Sawyer explained:* Interview with Sarah Sawyer, Yosemite National Park, Calif., July 14, 2021.

6: Canary in the Coal Mine

105 *"There he is":* Interview with Scott Stafford, Washington, D.C., March 4, 2022.

106 *thin refrain that sounds like:* "White-throated Sparrow Sounds," All About Birds website, https://www.allaboutbirds.org/guide/White-throated _Sparrow/sounds.

106 *"So I'm right by the window":* Interview with Scott Stafford, Washington, D.C., March 4, 2022.

107 *A couple from Sarasota, Florida, David Mann and Amy Donner:* Video interview with Haikubox originators David Mann and Amy Donner, June 23, 2022.

107 *"We believe that people ultimately care":* Video interview with Chad Wilsey, Nov. 14, 2021.

107 *Of all these tools, the project that has developed:* Series of interviews with Cornell Lab staff members over the course of 2021 and 2022.

108 *But Amanda Rodewald, the Cornell Lab's senior director:* Interview with Amanda Rodewald, Ithaca, N.Y., Aug. 31, 2021.

108 *"He asked if we shouldn't":* Interview with John Fitzpatrick, Vero Beach, Fla., Feb. 18, 2022.

108 *"We had no idea":* Video interview with Frank Gill, Feb. 22, 2022.

108 *In 1999, Cornell landed a $3 million National Science Foundation grant:* Video interview with John Fitzpatrick, July 9, 2022.

109 *"That next month it took off":* Video interview with Chris Wood, Jan. 27, 2022.

109 *The appeal was lasting, too:* Interviews with eBird leaders over 2022, in addition to data from the Cornell Lab website, https://ebird.org/home.

109 *"I can't begin to describe":* Interview with John Fitzpatrick, Ithaca, N.Y., Aug. 31, 2021.

110 *A century ago, more than four million acres:* Paige Blankenbuehler, "The Disappearing Wetlands of California's Central Valley," *High Country News*, Feb. 29, 2016.

110 *"But even after many, many years of that":* Interview with Mark Reynolds, March 6, 2019.

111 *The projects asked farmers to submit bids:* Interview with Greg Golet, Colusa County, Calif., May 19, 2021.

111 *The initial costs came to about:* Justine E. Hausheer, "Bumper-Crop Birds:

Pop-Up Wetlands Are a Success in California," Cool Green Science web-site, Jan. 29, 2018, https://blog.nature.org/science/2018/01/29/bumper -crop-birds-pop-up-wetlands-are-a-success-in-california/.

111 *"Now pretty much all the farmers have cards":* Interview with John Brennan, Colusa, Calif, May 19, 2021.

112 *"I wouldn't suggest this is the only way":* Video interview with Greg Golet, March 4, 2022.

112 *"They approached us to use eBird":* Interview with Viviana Ruiz-Gutierrez, Ithaca, N.Y., Sept. 2, 2021.

112 *"We're going to rely on eBird as much as we can":* Video interview with Jerome Ford, Oct. 14, 2021.

112 *The staff at Cornell expect to take eBird in new directions:* Interview with Ian Owens, Ithaca, N.Y., Aug. 30, 2021.

113 *"My hope is that in the next five years":* Ibid.

113 *Canaries were first put to work in coal mines:* Neil Prior, "How 1896 Tylor-stown Pit Disaster Prompted Safety Change," BBC News, Wales website, January 28, 2012, https://www.bbc.com/news/uk-wales-15965188.

113 *The birds stood sentinel until 1996:* "Singing as They Go, Miners' Little Friends Head for Retirement," *Daily Mail*, Jan. 2, 1996, https://link.gale.com/apps/doc/EE1860550990/GDCS?u=oxford&sid=GDCS&xid=1f1 e2b3a.

113 *invention of a handheld carbon dioxide monitor:* "Canaries in Coal Mines, End of an Era," Mining Heritage website, Feb. 7, 2020, https://mining heritage.co.uk/pit-canaries-end-of-an-era/.

113 *During the Persian Gulf War:* Claire Scobie, "Bye-Bye Birdie," (London) *Daily Telegraph*, Jan. 6, 1996.

113 *And after a 1995 terrorist gas attack in Tokyo:* Ibid.

114 *Recorders collect songs from Sonoma's 240 bird species:* "Soundscapes to Land-scapes," Soundscapes2landscapes.org, https://soundscapes2landscapes .org/.

114 *"What we're trying to see is the biodiversity on a landscape level":* Interview with Leo Salas, Sonoma County, Calif., May 21, 2021.

114 *Birds are also proving to be indicators:* Interview with Viviana Ruiz-Gutierrez, Ithaca, N.Y., Sept. 2, 2021.

114 *visited the Cornell Lab to see how birds could help gauge*: Ibid.

115 *"It's abundantly clear in all of this that if birds are doing well":* "When Is Birdsong the Sound of Sustainability," Nespresso website, https://www .sustainability.nespresso.com/birdsong-sound-of-sustainability.

115 *The project relies on eighty audio recorders:* Ibid.

115 *"This increases how much coffee they can generate":* Interview with Viviana Ruiz-Gutierrez, Ithaca, N.Y., Sept. 2, 2021.

115 *"I said, 'If you close your eyes'":* Ibid.

115 *"This is such an important emerging issue":* Video interview with Scott Sillett, Sept. 14, 2022.

116 *"We know many birds and pollinators respond similarly":* Marc Devokaitis, "What Can Birds Tell Us About Pollinators? Walmart-Funded Study Aims to Find Out," All About Birds website, Oct. 13, 2021, https://www.allaboutbirds.org/news/what-can-birds-tell-us-about-pollinators-walmart-funded-study-aims-to-find-out/.

116 *"I think groups like Walmart are really feeling":* Video interview with Ian Owens, June 20, 2022.

116 *"I think we're beginning to see more and more investment":* Video interview with Lynn Scarlett, March 1, 2022.

117 *Either way, Michael Doane, who travels the world:* Video interview with Michael Doane, April 1, 2022.

117 *"There's not a CEO in agribusiness":* Ibid.

117 *When another year of eBird information:* Interview with Scott Stafford, Washington, D.C., March 4, 2022.

117 *Peter Kaestner has logged as many species:* Video interview with Peter Kaestner, March 8, 2022.

118 *"I love the competition":* Ibid.

120 *A century ago, in the earliest days of wildlife photography:* Oliver Tatom, "William L. Finley, 1876–1953," *Oregon Encyclopedia,* https://www.oregonencyclopedia.org/articles/finley_william_l_1876_1953_/#.Yvppv OzMJ_R.

120 *These days,* National Geographic *photographer Joel Sartore:* "A Man on a Mission: Building the Photo Ark," Joel Sartore website, https://www.joelsartore.com/gallery/the-photo-ark/.

7: World Travelers

123 *By late May or early June, most of these:* Sasha Paris and Laura Erickson, "Declining Numbers of Cerulean Warblers," BirdScope, Cornell Lab of Ornithology, October 2009, https://www.allaboutbirds.org/news/declining-numbers-of-cerulean-warblers/.

123 *The industrious warblers then spend the next five days:* "Cerulean Warbler Life History," All About Birds website, https://www.allaboutbirds.org/guide/Cerulean_Warbler/lifehistory#.

124 *For the next couple of weeks, the female lays:* Ibid.

124 *In northern Ecuador, George Cruz is waiting:* Interview with George Cruz, Mindo, Ecuador, Jan. 9, 2022.

124 *"Such beautiful blue birds":* Ibid.

124 *North and South America share an estimated 350 long-distance migratory:* "The Basics of Bird Migration: How, Why, and Where," All About Birds website, Aug. 1, 2021.

124 *"This is a global problem":* Video interview with Paul Greenfield, Jan. 18, 2022.

124 *With 1,659 species, Ecuador has:* Garth C. Clifford, "Birds of Ecuador," World Birds: Joy of Nature, Aug. 4, 2021.

125 *they've lost population at the steepest rate:* "A Conservation Action Plan for the Cerulean Warbler (Dendroica cerulea)," U.S. Fish and Wildlife Service's Division of Migratory Bird Management Focal Species Program, June 30, 2007.

125 *Their numbers are still high enough for interventions*: "Cerulean Warbler Life History," All About Birds website, https://www.allaboutbirds.org/guide/Cerulean_Warbler/lifehistory#.

125 *If the current Cerulean habitat protections succeed on both continents:* Scott R. Loss, Tom Will, and Peter P. Marra, "Direct Mortality of Birds from Anthropogenic Causes," *The Annual Review of Ecology, Evolution, and Systematics* 46, no. 1 (December 2015): 99, https://www.annualreviews.org/doi/10.1146/annurev-ecolsys-112414-054133.

125 *Even though an impressive 20 percent of its land is protected":* Interviews with a series of conservation organizations in Ecuador put the amount of land under protection at about 20 percent. The World Bank's indicators place the portion at 23 percent. https://tradingeconomics.com/ecuador/terrestrial-protected-areas-percent-of-total-land-area-wb-data.html.

125 *On our third day here, Juan Carlos Crespo is the first to jump:* Interview with Juan Carlos Crespo, Mindo, Ecuador, Jan. 8, 2022.

126 *The conservancy has a clear aim:* Interview with Mike Parr, Mindo, Ecuador, Jan. 10, 2022.

127 *"It's like my head is exploding":* Interview with Juan Carlos Crespo, Mindo, Ecuador, Jan. 8, 2022.

128 *The trip shows off the unparalleled biodiversity:* "The Influence of Ice Ages and Ecotones on Biodiversity," Mongabay website, July 31, 2012, https://rainforests.mongabay.com/0304a.htm; "Flora and Fauna," Embassy of Ecuador in The Netherlands website, http://www.embassyecuador.eu/site/index.php/en/turismo-inf-general-2/turismo-flora-fauna.

128 *And the Cerulean's range is massive:* Sasha Paris and Laura Erickson,

"Declining Numbers of Cerulean Warblers," BirdScope, the Cornell Lab of Ornithology, October 2009; "Cerulean Warbler Life History," All About Birds website, https://www.allaboutbirds.org/guide/Cerulean_Warbler /lifehistory#.

128 *Roughly 80 percent of Ceruleans settle in the Appalachian Mountains*: Ibid.

128 *"We can safeguard the endangered endemics":* Interview with Mike Parr, Canandé, Ecuador, Jan. 12, 2022.

129 *This is where a lot of the benefits:* Interview with Todd Fearer, Blacksburg, Va., Sept. 8, 2021.

129 *the Jefferson National Forest is part of an $8 million conservation experiment*: Ibid.

130 *"A good 75 to 80 percent of this land":* Ibid.

130 *reimburse 75 percent of forest improvement costs:* Ibid.

130 *By that point the bird's losses were so precipitous*: "Cerulean Warbler Life History, All About Birds website, https://www.allaboutbirds.org/guide /Cerulean_Warbler/lifehistory.

131 *since 1966 Cerulean numbers had been declining by about 3 percent each year*: Paul B. Hamel, "Cerulean Warbler Status Assessment," U.S. Fish and Wildlife Service, April 2000; "Cerulean Warbler Status Assessment," U.S. Fish and Wildlife Service Midwest Region, April 2000, updated Jan. 2, 2020.

131 *"Honestly, there were a lot of state wildlife directors":* Video interview with Bob Ford, March 23, 2021.

131 *the Fish and Wildlife Service would need to move quickly:* "The Petition Process for Requests to List a Species as a Threatened or Endangered Species Under the Endangered Species Act," U.S. Fish and Wildlife Service Endangered Species Program, Aug. 2016, http://www.fws.gov /endangered/.

131 *The law gives the service fifteen months:* Ibid.

131 *A species makes the ESA list when it's in danger of extinction":* "Kristyn Judkins, "Deciphering the ESA's Enigmatic SPR Phrase," Environmental, Natural Resources & Energy Law blog, Lewis and Clark Law School website, https://www.lclark.edu/live/blogs/121-deciphering-the-esas-enigmatic -spr-phrase-.

131 *A "threatened" bird is "likely to become":* Ibid.

131 *When Fish and Wildlife denied the Cerulean protection in November 2006:* "12-Month Finding on a Petition to List the Cerulean Warbler (Dendroica cerulea) as Threatened with Critical Habitat," *Federal Register*, Dec. 6, 2006, https://www.federalregister.gov/documents/2006/12/06

/E6-20530/endangered-and-threatened-wildlife-and-plants-12-month
-finding-on-a-petition-to-list-the-cerulean.

131 *The service acknowledged*: Ibid.

132 *The Cerulean's denial came a full four years:* "The Petition Process for Requests to List a Species as a Threatened or Endangered Species Under the
Endangered Species Act."

132 *By 2011 the problem reached federal court*: Noah Greenwald, "Lawsuit
Launched to Speed Endangered Species Act Protection for 417 Species,"
Center for Biological Diversity press release, Aug. 23, 2016.

132 *but by the deadline, 417 species were still waiting*: Ibid.

132 *It has resorted to publishing a five-year work plan:* "National Listing Work-
plan," U.S. Fish and Wildlife Service website, https://www.fws.gov/proj
ect/national-listing-workplan.

132 *But even those predicted dates are often missed:* Ibid.

132 *in 2022 was to resolve the forty species with court-ordered deadlines:* Ibid.

132 *Over five years the group:* D. K. Dawson, T. B. Wigley, and P. D. Keyser,
"Cerulean Warbler Technical Group: Coordinating International Research and Conservation," *Ornithologia Neotropical* 23 (2012): 275.

132 *In 2007 the group laid out a roadmap:* "A Conservation Action Plan for the
Cerulean Warbler (Dendroica cerulea) produced for the USFWS Divi-
sion of Migratory Bird Management Focal Species Program," Technical
Group, Revised version, June 30, 2007.

132 *The group mostly pinpointed additional:* Ibid.

133 *Fearer thinks the results are sufficient:* Interview with Todd Fearer, Blacksburg, Va., Sept. 8, 2021.

133 *the warbler joins thousands of other birds:* "Choco-Darien-Western Ecuador: Choco-Manabi Conservation Corridor Briefing Book," Critical
Ecosystem Partnership Fund, Jan. 24, 2005, https://www.cepf.net/sites
/default/files/final.chocodarienwesternecuador.chocomanabi.briefing
book.pdf.

133 *he's taking us to see property he's evaluating for possible purchase*: Interview
with Martin Schaefer, Canandé, Ecuador, Jan. 12, 2022.

133 *Nineteen types of monkeys, jaguars:* Juan Freile, "Observation Guide of
Ecuadorian Primates in Nature Areas of Ecuador," ResearchGate, July
2011, https://www.researchgate.net/publication/260749146_Observa
tion_Guide_of_Ecuadorian_Primates_in_Natural_Areas_of_Ecuador.

133 *"I feel like I'm in a washing machine":* Interview with Dan Lebbin, Canandé,
Ecuador, Jan. 12, 2022.

134 *"Most days, we see twenty to forty of these":* Interview with Martin Schaefer, Canandé, Ecuador, Jan. 12, 2022.

134 *Once logging companies:* Ibid.

134 *"So everybody loses out"*: Ibid.

134 *"You can't do successful conservation":* Ibid.

135 *Only about 15 percent of the original rainforest:* "Ecuador Forest Figures," Mongabay, https://rainforests.mongabay.com/20ecuador.htm.

135 *Angel Paz was eking out a living:* Interview with Angel Paz, Nanegalito, Ecuador, Jan. 9, 2022.

136 *"I began to follow the bird":* Ibid.

136 *"There's a sort of snowballing thing":* Phone interview with Paul Greenfield, Jan. 18, 2022.

136 *called Refugio Paz de las Aves:* Interview with Angel Paz, Nanegalito, Ecuador, Jan. 9, 2022, along with material from the Le Refugio website, https://www.refugiopazdelasaves.com/.

136 *"We were paid $10":* Interview with Angel Paz, Nanegalito, Ecuador, Jan. 9, 2022.

137 *The bird park formula works:* Phone interview with Paul Greenfield, Jan. 18, 2022.

137 *"People think, 'I could do that.'":* Ibid.

137 *But that's starting to change:* "Philanthropies Pledge $5 Billion to 'Protecting Our Planet' Challenge," *Philanthropy News Digest,* Sept. 22, 2021, https://philanthropynewsdigest.org/news/philanthropies-pledge-5-billion-to-protecting-our-planet-challenge.

137 *the Earth Fund's first grants are projects*: "Bezos Earth Fund Invests $12 Million for Bird Conservation in Tropical Andes," National Audubon Society website, Dec. 6, 2021, https://www.audubon.org/news/bezos-earth-fund-invests-12-million-bird-conservation-tropical-andes#:~:text=Press%20Room,-Bezos%20Earth%20Fund%20Invests%20%2412%20Million%20in%20Bird%20Conservation%20in,Bird%20Conservancy%20and%20RedLAC%20members.

137 *The flow of conservation funds exceeds:* Video interview with Byron Swift, Jan. 31, 2022.

137 *"But this increase in funding":* Ibid.

138 *The National Audubon Society is leading a project:* Video interview with Aurelia Ramos, Jan. 26, 2022.

138 *His first assignment at Audubon:* Ibid.

138 *"We started with a blank sheet":* Ibid.

138 *"I see North American organizations":* Video interview with Itala Yépez, Dec. 9, 2021.

138 *"I remember early on in the U.S.":* Phone interview with Paul Greenfield, Jan. 18, 2022.

139 *"It's always been more difficult":* Video interview with Lynn Scarlett, March 1, 2022.

139 *"It should be $600 million":* Interview with Mike Parr, Canandé, Ecuador, Jan. 12, 2022.

139 *"If you have dollars going out of the U.S.":* Video interview with Byron Swift, Jan. 31, 2022.

140 *At thirty-eight feet above sea level, High Island:* "Hurricane Ike Impact on High Island, Texas," NASA Earth Observatory, https://earthobservatory .nasa.gov/images/9107/hurricane-ike-impact-on-high-island-texas.

141 *In 2021 at its Smith Oaks Sanctuary:* Paula Dittrick, "Walkway Spans High Island's Smith Oaks," Coastal Prairie Chapter, Texas Master Naturalist, April 1, 2021, https://txmn.org/coastal/walkway-traverses-smith-oaks-at -high-island/.

142 *The idea for the walkway came about:* Phone interview with Richard Gibbons, Houston Audubon's conservation director, April 27, 2021.

8: When All Else Fails

145 *Scientists will breed tens of millions of the male insects:* Interviews with members of the *Wolbachia* team in person and by video over the course of 2021 and 2022.

146 *Nearly 100 of Hawai'i's 140 native bird species:* "Paradise for Some— but an Ongoing Extinction Crisis for Birds," American Bird Conservancy website, https://abcbirds.org/program/hawaii/#:~:text=Since%20 humans%20arrived%2C%2095%20of,decades%20and%20are%20 likely%20extinct. Note: the exact number of extinctions varies somewhat with different studies, but there's no debate about the massive loss of native birds in Hawai'i amounting to about two thirds of the original native species. J. Michael Reed et al., "Long-term Persistence of Hawaii's Endangered Avifauna Through Conservation-Reliant Management," *Bioscience*, Oct. 1, 2012, https://academic.oup.com/bioscience/article/62 /10/881/238090#94384348.

146 *fifteen birds, including eight expected to be declared extinct:* Video interview with Chris Farmer, April 6, 2022.

146 *Formed 30 million years ago:* Erin Wayman, "What We're Still Learning About Hawaii," *Smithsonian Magazine*, December 2011, https://www

.smithsonianmag.com/travel/what-were-still-learning-about-hawaii
-74730/.

146 *In the thirty-six years since the Florida Dusky Seaside Sparrow:* Kim Steu-
termann Rogers, "Wave of Hawaiian Bird Extinctions Stresses the Islands' Conservation Crisis," National Audubon website, Oct. 6, 2021,
https://www.audubon.org/news/wave-hawaiian-bird-extinctions-stresses
-islands-conservation-crisis. Note: The estimated fifteen birds that have
gone extinct in the past three decades includes eight Hawaiian species the
U.S. Fish and Wildlife Service proposed removing from the Endangered
Species List due to extinction, although the final decision on those birds
is still pending.

146 *"Sometimes I feel like my job is like a hospice nurse":* Phone interview with
Justin Hite, June 8, 2021.

147 *"That ecosystem is not recoverable":* Video interview with Paul Schmidt,
Aug. 2, 2021.

147 *One funding assessment:* David L. Leonard Jr., "Recovery Expenditures
for Birds Listed Under the US Endangered Species Act: The Dispar-
ity Between Mainland and Hawaiian Taxa," *Biological Conservation,*
August 2008, https://www.sciencedirect.com/science/article/abs/pii/S000
6320708002085.

147 *"I think of Hawai'i as a":* Video interview with Noah Greenwald, Feb. 25,
2022.

148 *"Hawai'i is a small place":* Interview with Chris Farmer, Hawai'i, HI, May
27, 2021.

148 *The first time André Raine:* Interview with André Raine, Hanapepe,
Kaua'i, HI, June 4, 2021.

149 *"They asked him how he was":* Interview with Helen Raine, Hanapepe,
Kaua'i, HI, June 4, 2021.

149 *Both birds are shaped like tiny, winged dolphins:* "What's a seabird?," National Audubon Society website, https://ca.audubon.org/what-s-seabird.

149 *The seabirds come to Kaua'i and other islands:* Brett Hartl, "Federal Analysis: Kaua'i Power Lines Kill 1,800 Endangered Seabirds a Year," Center for Biological Diversity press release, Oct. 18, 2017, https://www
.biologicaldiversity.org/news/press_releases/2017/hawaiian-seabirds
-10-18-2017.php#:~:text=Federal%20Analysis%3A%20Kauai%20Power
%20Lines%20Kill%201%2C800%20Endangered%20Seabirds%20a%20
Year&text=LIHUE%2C%20Hawaii%E2%80%94%20A%20scientific
%20analysis,lines%20on%20Kauai%20every%20year.

150 *One detailed study found the Hawaiian Petrel had lost:* André Raine et al.,

"Declining Population Trends of Hawaiian Petrel and Newell's Shearwater on the Island of Kaua'i, Hawaii, USA," *The Condor*, Aug. 2017.

150 *"His knees are bad":* Video interview with Brad Keitt, March 29, 2022.

150 *had caused a long-running dispute:* Associated Press, "Kauai Utility: Guilty on Species Violations," *Honolulu Star Advertiser*, Dec. 2, 2010, https://www.staradvertiser.com/2010/12/02/breaking-news/kauai-utility-guilty-on-species-act-violations/amp/%7B%7Blink/.

151 *"The utility company was extremely obstructionist":* Phone interview with Brad Keitt, April 30, 2021. Note: The Kaua'i Island Utility Cooperative conceded there've been disagreements about the state of the island's seabirds over the years. But the utility insisted it has worked hard, investing more than $43 million in protection efforts over time. Communications Manager Beth Tokioka said the cooperative "has always taken very seriously its obligations" and sees significant progress in protecting the birds, particularly since the data on deaths and injuries improved. "We believe our efforts are working," she said.

151 *earlier federal analysis had reported 1,800 annual deaths:* Ibid.

151 *Birds hit the wires nearly 16,000 times:* Division of Forestry and Wildlife workshop, YouTube.com, June 3, 2021, https://www.youtube.com/watch?v=Jmm5YGq6qs8&t=9230s.

151 *The results have been dramatic:* Video interview with André Raine, July 19, 2022.

152 *the world's tallest mountain:* "Mauna Kea," National Geographic Resource Library website, https://media.nationalgeographic.org/assets/file/mauna kea-ngm-sept2012supp.pdf.

152 *The catch is that Mauna Kea:* Joe Phelan, "Is Everest Really the Tallest Mountain on Earth?," *LiveScience*, Dec. 26, 2021, https://www.livescience.com/tallest-mountain-on-earth.

152 *All along its sides:* Deborah Ward, "Early History of 'Island Ranching' in Hawaii is Theme of 2018 Historic Preservation Calendar," Hawaii.gov., Dec. 8, 2017.

153 *"This is one of the last big native dry forests":* Interview with Chris Farmer, Mauna Kea, Hawai'i, HI, May 27, 2021.

153 *The Palila rescue is far slower:* "Call to Action—Save the Palila!," Conservation Council for Hawaii newsletter, 2010, https://www.conservehawaii.org/wp-content/uploads/2019/11/CCH_Palila_ActionAlert.pdf.

153 *but now are limited to Mauna Kea:* "Important Bird Areas: Mauna Kea Mamane–Naio Forest," National Audubon Society website, https://www.audubon.org/important-bird-areas/mauna-kea-mamane-naio-forest.

153 *eating practically every part:* "Palila, Diet and Foraging," Cornell Lab of Ornithology Birds of the World website, https://birdsoftheworld.org /bow/species/palila/cur/foodhabits.

154 *Like the Palila's territory:* Interview with Jack Jeffrey, Mauna Kea, Hawai'i, HI, May 28, 2021.

155 *One day in the early 1990s:* Ibid.

156 *The realization that mosquitoes were killing off:* Carter T. Atkinson, with Dennis LaPointe, "Introduced Avian Diseases, Climate Change, and the Future of Hawaiian Honeycreepers," *Journal of Avian Medicine and Surgery*, May 1, 2009, https://bioone.org/journals/Journal-of -Avian-Medicine-and-Surgery/volume-23/issue-1/2008-059.1/Intro duced-Avian-Diseases-Climate-Change-and-the-Future-of-Hawaiian /10.1647/2008-059.1.short.

156 *Ten native species, most of them the island's:* Elizabeth Newbern, "How Malaria Hurts Birds," National Audubon Society website, June 16, 2015, https://www.audubon.org/news/how-malaria-hurts-birds.

156 *The native birds seemed able to survive only:* Ibid.

156 *Finally, in the mid-1960s:* U.S. Fish and Wildlife Service, Pacific Islands division, "What's Killing Hawaii's Forest Birds?," Medium website, Sept. 22, 2021, https://medium.com/usfwspacificislands/whats-killing -hawai%CA%BBi-s-forest-birds-dbd4cd2c33be.

156 *"If you read the old naturalists' accounts":* Video interview with Dennis LaPointe, March 14, 2022.

156 *Then a decade ago, the U.S. Geological Survey:* Interview with U.S. Geological Survey wildlife biologist Paul Banko, Volcano, HI, May 29, 2021.

157 *Mosquitoes buzzed steadily upward:* "Avian Disease: A Bane for Hawai'i's Forest Birds," Maui Forest Bird Recovery Project website, https://maui forestbirds.org/avian-disease/.

157 *she found herself battling the invasive southern house mosquito:* Katherine McClure, "Landscape-level Mosquito Suppression to Protect Hawaii's Rapidly Vanishing Avifauna," Cornell University Wildlife Health Center website, July 1, 2020, https://wildlife.cornell.edu/blog/land scape-level-mosquito-suppression-protect-hawaiis-rapidly-vanishing -avifauna.

157 *"I'm more worried than I've ever been":* Interview with Lisa "Cali" Crampton, Kaua'i, HI, June 3, 2021.

158 *First the mosquito bites the bird on the bare skin of its eye or legs:* Katherine McClure, "Landscape-Level Mosquito Suppression to Protect Hawaii's Rapidly Vanishing Avifauna."

158 *"My first year out, I found like fifteen nests":* Phone interview with Justin Hite, June 8, 2021.

159 *"It's our best hope":* Video interview with Teya Penniman, March 10, 2022.

159 *"I have a great deal of hope":* Interview with Sabra Kauka, Kalapaki Beach, HI, June 6, 2021.

160 *"When I'm being optimistic":* Video interview with Dennis LaPointe, March 14, 2022.

161 *This is the Frozen Zoo:* Interviews with Frozen Zoo leadership, via video and in person at the San Diego Zoo Wildlife Alliance offices, San Diego, Calif., May 10, 2021.

161 *"I imagine a future in which there's been tons":* Video interview with Beth Shapiro, May 17, 2021.

162 *"It's usually a false alarm":* Video interview with Marlys Houck, April 6, 2022.

162 *"I don't believe there's any place like this in the world":* Video interview with Cynthia Steiner, April 5, 2022.

162 *In 2020, a California-based nonprofit called Revive & Restore:* Rasha Aridi, "Scientists Cloned an Endangered Wild Horse Using the Decades-Old Frozen Cells from a Stallion," *Smithsonian Magazine,* Oct. 15, 2020, https://www.smithsonianmag.com/smart-news/save-endangered-wild -horse-species-scientists-cloned-stallion-using-its-decades-old-frozen -cells-180976069/.

162 *Then in 2021, Revive & Restore:* Sabrina Imbler, "Meet Elizabeth Ann, the First Cloned Black-Footed Ferret," *New York Times,* Feb. 18, 2021, https://www.nytimes.com/2021/02/18/science/black-footed-ferret -clone.html.

162 *One of the early genomic achievements:* Video interview with Cynthia Steiner, April 5, 2022.

162 *"We hope we don't have to use them":* Interview with Nadine Lamberski, San Diego, Calif., May 10, 2021.

163 *"We really want to figure out":* Video interview with Ryan Phelan, March 21, 2022.

163 *"This first room is where we'll incubate":* Interview with Ben Novak, Brevard, N.C., Nov. 29, 2021.

164 *A little more than a year ago:* Ibid.

164 *Stewart Brand, the ardent environmentalist:* "The Dawn of De-Extinction: Are You Ready," TED Talk, March 13, 2013, https://www.ted.com /talks/stewart_brand_the_dawn_of_de_extinction_are_you_ready ?language=en.

165 *The last Passenger Pigeon, a bird named Martha, died:* "Martha, the Last Passenger Pigeon," Smithsonian Museum of Natural History website, https://naturalhistory.si.edu/research/vertebrate-zoology/birds/collec tions-overview/martha-last-passenger-pigeon.

165 *The reception wasn't so warm:* David Shultz, "Bringing Extinct Species Back from the Dead Could Hurt—Not Help—Conservation Efforts," *Science,* Feb. 27, 2017, https://www.science.org/content/article/bringing -extinct-species-back-dead-could-hurt-not-help-conservation-efforts.

166 *"The problem is that we're being condemned":* Video interview with Beth Shapiro, May 17, 2021.

166 *The steps for bringing back the Passenger Pigeon:* Interview with Ben Novak, Brevard, N.C., Nov. 29, 2021; Revive & Restore website, https://re viverestore.org/about-the-passenger-pigeon/.

166 *Here's how Novak describes the mapping challenge:* Ibid.

167 *As Stewart Brand put it:* "The Dawn of De-Extinction: Are You Ready," TED Talk.

167 *"You come out somewhere like this":* Interview with Ben Novak, Brevard, N.C., Nov. 29, 2021.

168 *"So the choice is to lose everything":* Ibid.

169 *"I think in ten years' time":* Video interview with Ryan Phelan, March 21, 2022.

9: Coexisting with the Birds

171 *"I was probably six or seven":* Interview with Josh Hoy, Flint Hills, Kans., July 14, 2021.

173 *But as it turned out, the Hoys were merely ahead of the curve:* "Regenerative Agriculture Is Transforming Heifer Ranch into the Garden of Eden," The Cattle Site, Oct. 4, 2021, https://www.thecattlesite.com/news/57514 /regenerative-agriculture-is-transforming-heifer-ranch-into-the-gar den-of-eden/.

173 *Prairies historically covered nearly one third of North America:* "A Complex Prairie Ecosystem," National Park Service, Tallgrass Prairie National Pre serve Kansas website, https://www.nps.gov/tapr/learn/nature/a-complex -prairie-ecosystem.htm.

173 *across half a billion acres:* C. B. Wilsey et al., "North American Grasslands," National Audubon Society, New York, N.Y., 2019. https://www.audubon .org/conservation/working-lands/grasslands-report.

173 *home to 450 different types of grasses:* James Stubbendieck et al., "Grasses of the Great Plains," Texas A&M University Press website, Feb. 17, 2017,

https://www.tamupress.com/book/9781623494773/grasses-of-the-great
-plains/.

173 *the United States has lost more prairie than the Brazilian Amazon has
rainforest*: Chelsea Harvey, "North America's Grasslands Are Slowly
Disappearing—and No One's Paying Attention," *Washington Post*,
Nov. 29, 2016.

173 *Most Americans have no idea of the dimensions:* Video interview with Marshall Johnson, March 4, 2021.

173 *The deep, perennial root systems of these native plants*: Wilsey et al., "North
American Grasslands."

174 *The Three Billion Bird research showed grassland birds as a whole*: Kenneth V. Rosenberg et al., "Decline of the North American Avifauna," *Science*, Sept. 19, 2019.

174 *Eastern Meadowlarks, the voice of the grasslands:* Gustave Axelson, "Scientists Use the 'Half-Life' of a Species to Motivate Conservation," *LivingBird
Magazine*, Spring 2017, https://www.allaboutbirds.org/news/scientists
-use-the-half-life-of-a-species-to-motivate-conservation-efforts/.

174 *The Greater Sage-Grouse, the symbol of the West:* Andy McGlashen, "Greater
Sage-Grouse Has Plunged by 80 Percent Since 1965," National Audubon Society website, March 30, 2021, https://www.audubon.org/news
/greater-sage-grouse-populations-have-plunged-80-percent-1965.

175 *It all started when Josh was speeding along*: Interview with Josh, Gwen, and
Josie Hoy, Flint Hills, Kans., July 14, 2021.

176 *"I thought, well, if I can't cowboy:* Ibid.

176 *The last 4 percent of the tallgrass prairie:* "Last Stand of the Tallgrass Prairie," National Park Service, Tallgrass Prairie National Preserve website,
https://www.nps.gov/tapr/index.htm.

176 *where a unique geology of limestone and shale makes the soil too rocky to plow:*
"Tallgrass Prairie: Geology in the Flint Hills," National Park Service,
Tallgrass Prairie National Preserve website, https://www.nps.gov/tapr
/learn/nature/geology-at-the-preserve.htm.

176 *While 80 percent of a tallgrass prairie's plants are grasses*: "A Complex Prairie Ecosystem," National Park Service, Tallgrass Prairie National Preserve
Kansas website. https://www.nps.gov/tapr/learn/nature/a-complex-prairie
-ecosystem.htm.

176 *like bluestem, asters, and lupine, cover the rest:* "Tallgrass Prairie Ecosystem," LandScope America website., http://www.landscope.org/explore
/ecosystems/disappearing_landscapes/tallgrass_prairie/.

176 *Aldo Leopold, the naturalist and writer considered the father of modern*

conservation: "Aldo Leopold," Aldo Leopold Foundation website, https://www.aldoleopold.org/about/aldo-leopold/.

177 *Descendants of the Corriente cattle that can be traced back:* "Breeds of Livestock—Corriente Cattle," Ohio State University, Department of Animal Science.

178 *Three years ago, they won the Kansas Aldo Leopold Conservation Award:* "Hoy Family Flying W Receives Kansas Aldo Leopold Conservation Award," Sand County Foundation website, Nov. 24, 2020.

179 *The idea is that ranchers sell their beef at a premium of up to 70 percent:* Video interview with Marshall Johnson, April 26, 2021.

179 *"We felt that one of the critical pieces that was missing":* Video interview with Marshall Johnson, June 2, 2022.

179 *A few years ago, the Hoys went into partnership with the Bobolink Foundation:* Video interview with foundation chairman Wendy Paulson and chief conservation officer Justin Pepper, Jan. 27, 2021.

179 *"There was no hint of those sparrows out there before":* Phone interview with Justin Pepper, May 7, 2022.

180 *No bird in North America puts on a better show:* M. A. Schroeder, "Female Eventually Solicits Copulation," "Greater Sage-Grouse (*Centrocercus urophasianus*)," A. F. Poole and F. B. Gill, eds., *Birds of the World* (Ithaca, N.Y.: Cornell Lab of Ornithology, version 1.0), https://doi.org/10.2173/bow.saggro.01.

180 *Each of the last three White House administrations swung back and forth:* "Sage Grouse Has Been a Political Football for Decades," Associated Press, Nov. 21, 2021. Note: For years, federal and state agencies, western politicians, and industry and nonprofit organizations have fought over the status of the Greater Sage-Grouse as its population gradually decreased. Throughout, the U.S. Fish and Wildlife Service has held off listing the bird under the Endangered Species Act although the species qualified for protection based on its losses. A compromise in 2015 worked out by a cross section of interest groups settled on a middle ground that enhanced protection short of a listing for the bird without curtailing industry or agriculture. The Trump administration threw out the agreement in 2017, a decision that was then reversed in court. The Biden administration has signaled plans to tighten protections for the birds, but has yet to act. That has made voluntary efforts such as the Sage Grouse Initiative all the more important since political wavering shows no signs of ending.

181 *They called it the Sage Grouse Initiative:* Video interviews with Dave Naugle, July 1, 2021, and May 7, 2022.

181   *Nevertheless, the Sage Grouse Initiative has grown into*: Ibid.

181   *The project comes with a sizable bill:* "Sage-Grouse Spending to Top $750 Million by 2018," Associated Press, Feb.13, 2015; video interview with Dave Naugle, May 7, 2022.

182   *"Birders, ranchers, the government, energy companies":* Video interview with Tim Griffith, Oct. 20, 2021.

182   *"I tend to think trying to strike some balance":* Interview with Brian Jensen, Pinedale, Wyo., July 6, 2021.

182   *"I think most ranchers are happy to coexist":* Interview with Madeleine Murdock, Pinedale, Wyo., July 7, 2021.

184   *"I love it because, here":* Interview with Hilary Swain, Lake Placid, Fla., Feb. 17, 2021.

185   *Buck Island's success is convincing in itself:* Ibid.

185   *"A conservation strategy that just protects public lands will never succeed":* Video interview with Hilary Swain, Aug. 4, 2022.

185   *In the United States, slightly more than 60 percent of the lands:* "Public and Private Land Percentages by U.S. States," Summitpost.org., https://www.summitpost.org/public-and-private-land-percentages-by-us-states/186111.

186   *"Traditionally, people thought conservation meant":* Interview with Amanda Rodewald, Ithaca, N.Y., Aug. 31, 2021.

186   *"It's absolutely alive":* Video interview with Michael Doane, April 1, 2022.

186   *"You hear a healthy prairie before you see it or smell it":* Interview with Josh Hoy, July 15, 2021.

10: Case Studies in Getting It Done

191   *"I think we do about 2,500 of these across the country":* Interview with Will Johnson, Raleigh, N.C., Nov. 4, 2021.

192   *Ducks Unlimited banquets and tournaments are fundraising forces:* Interviews with Will Johnson and Ducks Unlimited documents.

192   *But tonight's crowd that will contribute a record $187,000:* Ibid.

192   *"You learn this from your father":* Interview with Bennett Whitehouse, Raleigh, N.C., Nov. 4, 2021.

193   *The nonprofit has 700,000 members*: Interviews with Ducks Unlimited leadership; Michael Furtman, "The Ducks Unlimited Story: Conservation for Generations," Ducks Unlimited, Memphis, Tenn., Nov. 2011.

193   *"I could call five to ten billionaires":* Video interview with Paul Schmidt, Aug. 2, 2021.

193   *Since its passage in 1937:* Lisa Irby, "Celebrating 80 Years of the

Pittman-Robertson Act," Ducks Unlimited website, https://www.ducks
.org/conservation/public-policy/celebrating-80-years-of-the-pittman
-robertson-act.

193 *Waterfowl are the most watched bird group in the country:* "Birding in the
United States: A Demographic and Economic Analysis Addendum to
the 2006 National Survey of Fishing, Hunting, and Wildlife-Associated
Recreation," U.S. Fish and Wildlife Service publications, 2009, http://
digitalcommons.unl.edu/usfwspubs/164.

193 *Game bird populations rose 56 percent:* Kenneth V. Rosenberg et al., "Decline of the North American Avifauna," *Science,* Sept. 19, 2019, https://
pubmed.ncbi.nlm.nih.gov/31604313/.

193 *"I don't ever miss the opportunity":* Video interview with Adam Putnam,
Aug. 23, 2021.

194 *"The Audubon folks, the Nature Conservancy folks":* Interview with Brian
Madison, Raleigh, N.C., Nov. 4, 2021.

195 *Richard Williams started guiding hunters as a teenager six decades ago:* Interview with Richard Williams, Knotts Island, N.C., Jan. 19, 2021.

195 *Its geology created a rare ecosystem:* Frank Tursi, "Duck Dynasty: When
Waterfowl Ruled the Roost," CoastalReview.org, March 11, 2014,
https://coastalreview.org/2014/03/duck-dynasty-when-waterfowl-ruled
-the-roost/.

195 *filling the bay with mostly freshwater:* "Currituck Sound Coalition Marsh
Conservation Plan," Audubon North Carolina, December 2021, https://
nc.audubon.org/sites/default/files/static_pages/attachments/currittuck
_sound_marsh_conservation_plan_202109_final_2.pdf.

195 *"Ducks and geese follow in numbers that stagger the imagination":* Tursi,
"Duck Dynasty: When Waterfowl Ruled the Roost."

195 *They included the head of U.S. Steel:* Ibid.

195 *Presidents Herbert Hoover and Dwight Eisenhower:* "Knotts Island, N.C.:
History, Beauty and Community," ThinkCurrituck.com, https://www
.thinkcurrituck.com/blog/knotts-island-nc-history-beauty-community.

195 *So when the game populations began to disappear:* T. Edward Nickens and
Peter Frank Edwards, "Knapp's Island: A Duck Hunter Travels to the
Spiritual Home of Ducks Unlimited," https://www.ducks.org/hunting
/waterfowl-hunting-destinations/knapps-island.

195 *Williams was just a boy:* Interview with Richard Williams, Knotts Island,
N.C., Jan. 19, 2021.

196 *Knapp ran the New York publishing house:* "Joseph P. Knapp, Publisher,
Is Dead," *New York Times,* Jan. 31, 1951; T. Edward Nickens and Peter

Frank Edwards, "Knapp's Island: A Duck Hunter Travels to the Spiritual Home of Ducks Unlimited," *Ducks Unlimited*, September/October 2021; https://www.ducks.org/hunting/waterfowl-hunting-destinations/knapps-island.

196 *DU was formally launched in upstate New York*: Ibid.

196 *"My Lord. He was the backbone of the community"*: Phone interview with Richard Williams, June 7, 2022.

197 *Though populations rebounded*: Adam Putnam, "NAWCA: A Proven Model for Saving North America's Birds," North American Waterfowl Management Plan website, Feb. 25, 2020, https://nawmp.org/node/320.

197 *The preserves are more important than ever now*: Kate Mosher, "Currituck Sound Coalition Announces Conservation Plan for Critical Marshes Adapting to Climate Change," *Sea Grant North Carolina News*, November 30, 2021, https://ncseagrant.ncsu.edu/news/2021/11/currituck-sound-coalition-announces-conservation-plan-for-critical-marshes-adapting-to-climate-change/.

198 *"Hunters and anglers are frankly the primary funders"*: Video interview with Adam Putnam, Aug. 23, 2021.

198 *"Ninety percent of the money"*: Interview with Pete Marra, Ithaca, N.Y. Aug. 31, 2021.

198 *"That's because the history and the foundation"*: Video interview with Adam Putnam, Aug. 23, 2021.

198 *an estimated $4 billion in birdseed sold each year*: Todd Whitesel, "Americans Spend More than $75 Billion Watching Wildlife," Realtree.com, April 25, 2018, https://business.realtree.com/business-blog/americans-spend-more-75-billion-annually-watching-wildlife.

198 *Birding enthusiasts tried at one point to increase dedicated funding*: Tom Stienstra, "New Equipment Tax Would Help Wildlife and Conservation," *San Francisco Examiner*, https://www.sfgate.com/sports/article/New-equipment-tax-would-help-wildlife-3135099.php. John Goodell, "Americans Love Public Lands and Species Conservation but How Do We Pay for Them?," *Mountain Journal*, Sept. 26, 2019, https://mountainjournal.org/wildlife-conservation-gets-a-game-changing-bill-in-congress.

198 *The stamp's illustrations are chosen*: "Federal Duck Stamp," U.S. Fish and Wildlife Service website, https://www.fws.gov/program/federal-duck-stamp/federal-duck-stamp-contest-event-information.

198 *the stamps typically become collectibles*: "2020–2021 Migratory Bird and Conservation Stamp: A Unique and Collectible Work of Art," U.S. Fish and Wildlife Service, June 29, 2020.

198 *"I thought, let's float this idea":* Video interview with Paul Schmidt, Aug. 2, 2021.

199 *"Who knows how many hundreds of thousands":* Video interview with Will Johnson, May 11, 2022.

199 *Not only was the proposal rebuffed:* Andy McGlashen, "After Controversy, the Duck Stamp Contest May No Longer Require Hunting Scenes," National Audubon Society website, June 16, 2021.

200 *"I think they're perfect partners":* Interview with Ashley Dayer, Blacksburg, Va., Sept. 7, 2021.

200 *When a cross section of representatives:* Paul A. Smith, "'Game Changing' Bill for Fish and Wildlife Agencies in America Moves Closer to Passing," *Milwaukee Journal*, April 10, 2022.

200 *A blue-ribbon panel, led by former Wyoming governor David Freudenthal:* Video interview with David Freudenthal, May 26, 2022.

200 *"I think this is a noble effort":* Ibid.

201 *trains more than half of the nation's new recruits*: Tracy Fuga, "Fort Benning Georgia: In-Depth Welcome Center (2022 Edition)"; *My Base Guide*, Fort Benning Community website, https://mybaseguide.com/installa tion/fort-benning/community/fort-benning-georgia-welcome-center/.

201 *has been forced to play nursemaid to more endangered species:* Bruce A. Stein, Cameron Scott, and Nancy Benton, "Federal Lands and Endangered Species: The Role of Military and Other Federal Lands in Sustaining Biodiversity," *BioScience* 58, no. 4 (April 1, 2008), https://doi.org/10.1641 /B580409.

202 *At its Seal Beach weapons station:* Benjamin Landis. "Sediment Strategy Seeks to Save Salt Marsh Species," U.S. Geological Survey, U.S. Climate Resiliency Toolkit, 2017, https://toolkit.climate.gov/case-studies/sedi ment-strategy-seeks-save-salt-marsh-species.

202 *At Utah's Dugway Proving Ground:* Amy Joi O'Donoghue, "Saving the Golden Eagle," *Deseret News*, July 29, 2021.

202 *The army is installing bioacoustic monitors:* Interview with Cornell Lab's Holger Klinck, Ithaca, N.Y., Sept. 1, 2021.

202 *biologist John Macey puts geolocating backpacks on the warblers:* Interview with John Macey, Killeen, Tex., May 3, 2021.

202 *spread throughout fifteen southeastern military bases*: David K. Delaney et al., "Response of Red-Cockaded Woodpeckers to Military Training Op-erations," *Wildlife Monographs*, July 2011, https://wildlife.onlinelibrary .wiley.com/doi/10.1002/wmon.3.

202 *just minutes later, 4.4 to be exact:* Ibid.

202 *"A lot of people think, 'Oh, it's endangered'":* Interview with James Parker, Columbus, Ga., April 7, 2021.

202 *Fort Benning was bulldozing through:* Tim W. Clark, Richard P. Reading, and Alice L. Clarke, eds., *Endangered Species Recovery: Finding the Lessons, Improving the Process* (Washington, D.C.: Island Press, 1994), 167.

203 *"I ain't never seen a red-cockadoodled woodpecker":* Ibid.

203 *A federal grand jury indicted three civilian employees:* Interview with James Parker, Columbus, Ga., April 7, 2021.

203 *"They had a hiccup":* Ibid.

203 *Training officers were caught treating:* David N. Diner, "The Army and the Endangered Species Act: Who's Endangering Whom?," Thesis Presented to the Judge Advocate General's School, United States, Defense Technical Information Center website, April 1, 1993, https://apps.dtic.mil/sti/citations/ADA456541.

203 *at the largest military base in the country:* Everett Bledsoe, "Top 5 Largest Military Bases in the World by Population & Area," The Soldiers Project, May 2, 2022, https://www.thesoldiersproject.org/largest-military-bases-in-the-world/.

203 *"Across the army as a whole, it was kind of the starting point":* Interview with Tim Marsden, Columbus, Ga., April 7, 2021.

203 *The Department of Defense now has two separate research departments:* Office of the Under Secretary of Defense, Defense Budget Overview, Fiscal Year 2023, https://comptroller.defense.gov/Portals/45/Documents/defbudget/FY2023/FY2023_Budget_Request_Overview_Book.pdf.

203 *the underlying reason for all its environmental work:* "The Strategic Environmental Research and Development Program (SERDP) and the Environmental Security Technology Certification Program (ESTCP) Resource Conservation and Resiliency (RCR) Program Area Research Plan," U.S. Department of Defense, Feb. 2021.

204 *"The Department of Defense is not a conservation organization":* Video interview with Ryan Orndorff, June 14, 2022.

204 *Every three years, foresters usher these fires:* Interview with James Parker, Columbus, Ga., April 7, 2021.

204 *One of Parker's staff had to be airlifted:* Ibid.

204 *"It flashed and burnt my whole face":* Ibid.

204 *vast Longleaf forests that once covered 92 million acres:* Cecil C. Frost, "Four Centuries of Changing Landscape Patterns in the Longleaf Pine Ecosystem," North Carolina Plant Conservation Program, Plant Industry Division, North Carolina Department of Agriculture, 1995.

204 *the tree is of no use to the woodpecker until it's at least eighty to a hundred years old*: David K. Delaney et al., "Response of Red-Cockaded Woodpeckers to Military Training Operations," *Wildlife Monographs*, July 2011, https://wildlife.onlinelibrary.wiley.com/doi/10.1002/wmon.3.

204 *The bird's main predator—the rat snake*: "Recovery Plan for the Red-Cockaded Woodpecker," U.S. Fish and Wildlife Service Southeast Region, Jan. 27, 2003.

205 *The solution came from a North Carolina State University professor*: Interview with Jeff Walters, Blacksburg, Va., Sept. 9, 2021.

205 *dubbed the "woodpecker nation"*: Ibid.

205 *In the late 1980s, they decided*: Ibid.

206 *"It works," she yelled over the payphone. "It actually works."*: Ibid.

206 *in September of 1989. In all, Hugo killed dozens of people and caused some $10 billion in damage*: "Hurricane Hugo," National Weather Service website, Sept. 16, 2014, https://www.weather.gov/ilm/hurricanehugo.

206 *"They lost 90 percent"*: Interview with Jeff Walters, Blacksburg, Va., Sept. 9, 2021.

206 *For decades, the base has planted*: Interview with James Parker, Columbus, Ga., April 7, 2021.

207 *In the earliest days of its life*: Jerome A. Jackson, "Red-Cockaded Woodpecker," Cornell Lab of Ornithology Birds of the World website, Jan. 1, 1994, https://birdsoftheworld.org/bow/species/recwoo/cur/.

207 *By the time the training was over five hours later*: Interview with James Parker, Columbus, Ga., April 7, 2021.

207 *"Anytime I have a dead Red-cockaded Woodpecker"*: Ibid.

208 *top administration officials were out in force*: "Trump Administration Proposes Downlisting of Red-Cockaded Woodpecker Under Endangered Species Act," U.S. Fish and Wildlife Service press release, Sept. 25, 2020.

209 *"This is a tremendous success story"*: Ibid.

209 *Two years of haggling over the specific*: Interview with Jeff Walters, Blacksburg, Va., Sept. 9, 2021.

209 *In fiscal year 2017 alone*: "Federal and State Endangered and Threatened Species Expenditures Fiscal Year 2017," U.S. Fish and Wildlife Service, 2017.

209 *But Fort Benning will continue*: Interview with Jeff Walters, Blacksburg, Va., Sept. 9, 2021.

209 *and is twenty-five years ahead of schedule*: "Recovery Plan for the Red-Cockaded Woodpecker," U.S. Fish and Wildlife Service Southeast Region, Jan. 27, 2003.

209 *Orndorff, the Department of Defense's head*: Video interview with Ryan Orndorff, June 14, 2022.

209 *"That's what the Red-cockaded Woodpecker"*: Ibid.

210 *Two months after the publication*: This account is based on a series of interviews with Pete Marra, Ken Rosenberg, and Tom Will over the course of 2021 and 2022.

210 *"You don't just go back to your day job"*: Ibid.

210 *"There was this explosion of ideas"*: Interview with Tom Will, Washington, D.C., Oct. 10, 2021.

211 *"What's clear to me is something is not working"*: Interview with Pete Marra, Washington, D.C., Oct. 12, 2021.

211 *"We're constantly hitting our heads against the wall"*: Interview with Ashley Dayer, Blacksburg, Va., Sept. 7, 2021.

211 *"The dirty little secret"*: Interview with Pete Marra, Washington, D.C., Oct. 12, 2021.

212 *"National Audubon could have done that, but they didn't"*: Ibid.

212 *"I don't get excited about the numbers"*: Video interview with Jerome Ford, Oct. 14, 2021.

212 *"They've been really beaten down"*: Video interview with Jamie Rappaport Clark, June 9, 2022.

212 *"The unfortunate thing is that the Fish and Wildlife"*: Video interview with Don Barry, June 3, 2022.

213 *The most detailed study of the Endangered Species recovery spending:* Noah Greenwald et al., "Shortchanged: Funding Needed to Save America's Most Endangered Species," Center for Biological Diversity website, 2017, https://www.biologicaldiversity.org/programs/biodiversity/pdfs/Shortchanged.pdf.

213 *"The Endangered Species Act is not failing"*: Video interview with Jamie Rappaport Clark, June 9, 2022.

213 *"Why aren't we making better arguments"*: Interview with John Fitzpatrick, Ithaca, N.Y., Aug. 31, 2021.

213 *"They're a shadow of their former selves"*: Video interview with Don Ashe, March 17, 2022.

214 *"It's driven by the need to publish"*: Interview with Mike Parr, Mindo, Ecuador, Jan. 14, 2022.

214 *"It's just not happening"*: Video interview with Dave Naugle, July 1, 2021.

214 *Road to Recovery has succeeded in its initial goal:* A series of video and phone interviews with Paul Schmidt over 2021 and 2022.

215 *"There's a tremendous thirst"*: Video interview with Paul Schmidt, Sept. 9, 2022.

215 *"We're hoping this will build enormous buzz":* Interview with Pete Marra, Washington, D.C., Oct. 12, 2021.

Conclusion: Making the Case for Birds

217 *We name our sports teams for birds:* Massey Ratings, a sport statistician website, MasseyRatings.com., https://masseyratings.com/mascots.

217 *Mahalia Jackson singing the cherished refrain:* "His Eye is on the Sparrow, A Brief Song History," Country Thang Daily website, https://www.countrythangdaily.com/eye-sparrow-grammy-jackson/.

217 *A Dutch painting,* The Goldfinch: "The Goldfinch, Donna Tartt (Little, Brown)," Pulitzer.org., https://www.pulitzer.org/winners/donna-tartt.

217 *This 1654 painting by Carel Fabritius:* "The Goldfinch," Mauritshuis website, https://www.mauritshuis.nl/en/our-collection/artworks/605-the-goldfinch/.

217 *Emily Dickinson's 1818 poem:* Emily Dickinson, "'Hope' Is the Thing with Feathers," *The Complete Poems of Emily Dickinson,* Thomas H. Johnson, ed. (Cambridge: The Belknap Press of Harvard University Press, 1951), https://poemanalysis.com/emily-dickinson/hope-is-the-thing-with-feathers/.

218 *Even today aerospace engineers study the mechanics:* Mark Phelps, "New Research Suggests There's Still a Lot to Learn from Birds," *AVweb,* May 17, 2022, https://www.avweb.com/aviation-news/new-research-suggests-theres-still-a-lot-to-learn-from-birds/.

218 *Studies of gulls and hummingbirds:* Liz Do, "A Better Way to Fly? U of T and UBC Researchers Study Birds and Their Wings," *University of Toronto News,* Jan. 7, 2019, https://www.utoronto.ca/news/better-way-fly-u-t-and-ubc-researchers-study-birds-and-their-wings.

218 *and hummingbirds guide engineers toward more efficient drones and robots:* Lakshmi Supriya, "New Study on Hummingbird Flight Reveals Surprising Twists," *The Wire,* Feb. 19, 2018, https://thewire.in/science/new-study-hummingbird-flight-reveals-surprising-twists.

218 *the global value of wildlife biodiversity:* "How much is nature worth? $125 trillion, according to this report," World Economic Forum, Oct. 30, 2018, https://www.weforum.org/agenda/2018/10/this-is-why-putting-a-price-on-the-value-of-nature-could-help-theenvironment/.

218 *More than 8,000 species of plants and flowers:* "Hummingbird Pollination Practice," National Audubon Society website, April 22, 2022, https://www.audubon.org/news/hummingbird-pollination-practice.

218 *One Dutch study found that where birds are around:* Cagan H. Sekercioglu, Daniel G. Wenny, and Christopher J. Whelan, eds., *Why Birds Matter:*

Avian Ecological Function and Ecosystem Services (Chicago: University of Chicago Press, 2016.), VIIII.

218 *About 20 percent of bird species also spread seeds:* Ibid., 76.

218 *On Hawai'i's Kaua'i:* Phone interview with Lisa Crampton, Nov. 5, 2019.

219 *"We have no forest without the Puaiohi":* Ibid.

219 *Hawai'i has already experienced:* Jamie K. Reaser, "Agreements, International," *Encyclopedia of Biological Invasions*, edited by Daniel Simberloff and Marcel Rejmanek (Berkeley: University of California Press, 2011), 4–7, https://doi.org/10.1525/9780520948433-006.

219 *In one infamous case:* Meir Rinde, "Poison Pill: The Mysterious Die-off of India Vultures," Science History Institute, Sept. 3, 2019.

219 *with an overall economic impact:* Sekercioglu et al., *Why Birds Matter*, x.

219 *"Until one of these birds starts to go extinct":* Interview with Erin Katzner, Boise, Idaho, June 21, 2021.

219 *"They're among nature's masterpieces":* Interviews with John Fitzpatrick, Lake Placid, Fla., Feb. 16, 2021, and Ithaca, N.Y., Aug. 31, 2021.

220 *After traveling the equivalent of the circumference of the globe:* Note: We traveled most of 2021 and into 2022. We spent the first seven months roaming the continent, including stops in coastal North Carolina before making our way down the East Coast to Central and South Florida for several weeks. We then headed back up to Georgia to visit Fort Benning and across the Gulf Coast states. We spent time along the coast of Louisiana and Texas during the spring migration, and then drove along the Southwest to California, making stops in San Diego, Sacramento, and several Northern California locations. We parked the Airstream in Sacramento to fly to Hawai'i for several weeks, and then returned to California to travel the Sierra Nevada range. From there, we headed back east across Nevada, Oregon, Idaho, and into Wyoming. We then gradually worked our way southeast with stops in Nebraska, Kansas, and Tennessee before reaching home in Raleigh in late July. We headed back out for a northeastern swing that included visits to the Cornell Lab in upstate New York, Pittsburgh, and Blacksburg, Virginia. In the fall, we traveled to Washington and Maryland and took several trips around the Southeast. In early 2022, we flew to Ecuador and also took an Airstream trip back to Florida before returning home in the late spring.

220 *"Birds are telling us":* Video interview with Elizabeth Gray, Sept. 16, 2021.

222 *"We're not saying these birds are going extinct":* Interview with Ken Rosenberg, Ithaca, N.Y., Aug. 30, 2021.

222 *faces threats from poisoning caused by lead bullets":* Vincent A. Slabe et al.,

"Demographic Implications of Lead Poisoning for Eagles Across North America, *Science*, Feb. 17, 2022, https://www.science.org/doi/10.1126/science.abj3068#:~:text=Frequency%20of%20lead%20poisoning%20was,(0.7%25%2C%200.9%25).

222 *But the agency has declined to use its authority:* Video interview with Michael Bean, former assistant secretary at the U.S. Fish and Wildlife Service and legal expert on the Endangered Species Act. June 9, 2022. Note: Bean said of a potential lead bullet ban: "My biggest regret from the seven years I spent at Interior was my inability to make much headway on this issue. I think the source of this mortality is entirely avoidable. The service has the authority and should use it."

222 *These failures symbolize the inadequate current policies:* This assessment of the U.S. Fish and Wildlife Service performance in the wake of the loss of a third of the breeding birds draws on dozens of interviews with those who have worked closely with and on the staff of the agency over the years. The service deserves credit for its decades of work in protecting birds and other wildlife and for so far largely keeping extinctions to a minimum except in Hawai'i. But the agency is not gearing up for what clearly is a new era that calls for more aggressive responses—and as a rule has made no effort to explain itself. We spent a year requesting interviews with current leadership, almost all of which were turned down. We submitted lists of questions on various cases, only one of which elicited a response. We put in repeated requests to talk with Martha Williams, the agency's director, and Gary Frazer, who oversee the Endangered Species section. Frazer agreed to talk with us and then changed his mind. Williams's office agreed to arrange an interview, but eventually said her schedule didn't permit it. We were able to talk with more than a dozen current and former Fish and Wildlife leaders, including former directors. Several of these veterans said the political pressure on the agency makes leadership hesitant to conduct interviews. But that has meant the agency's voice and perspective is almost entirely missing at a time when its work is especially important.

222 *Even insiders who've devoted their careers:* Video interview with Don Barry, June 3, 2022.

223 *"They just don't have the depth of field":* Video interview with Dan Ashe, March 17, 2022.

223 *The legislation itself hasn't been reauthorized:* "Reauthorization of the Endangered Species Act: A Comparison of Pending Bills and a Proposed Amendment with Current Law," EveryCRSreport.com, March 13, 2006, https://www.everycrsreport.com/reports/RL33309.html.

223 *On average, rulings on whether to add species to the ESA list take twelve years:* Shreya Dasgupta, "Endangered Species Often Wait 12 Years or More for Protection," Mongabay, Aug. 25, 2016, https://news.mongabay .com/2016/08/endangered-species-often-wait-12-years-or-more-to-be -listed-for-protection/.

223 *More than fifty species of wildlife and plants have gone extinct:* Noah Greenwald et al., "Shortchanged: Funding Needed to Save America's Most Endangered Species," Center for Biological Diversity website, 2017, https:// www.biologicaldiversity.org/programs/biodiversity/pdfs/Shortchanged .pdf.

224 *Despite extensive development along coasts and waterways:* Kenneth V. Rosenberg et al., "Decline of the North American Avifauna," *Science*, Sept. 19, 2019.

224 *Ryan Orndorff, director of natural resources for the Department of Defense:* Video interview with Ryan Orndorff, June 14, 2022.

224 *"Conservation is going to be extraordinarily dependent on working farms:* Video interview with Hilary Swain, Aug. 4, 2022.

224 *We'll spend $40 billion for a repeat trip to the moon:* Michael Sheetz, "Here's What's at Stake for NASA's Artemis I Mission to the Moon," CNBC, Aug. 27, 2022, https://www.cnbc.com/2022/08/27/nasas-artemis-1-mis sion-what-you-should-know-about-sls-orion.html.

225 *We invest $200 billion annually:* "Highway and Road Expenditures," Urban institute website, https://www.urban.org/policy-centers/cross-center-ini tiatives/state-and-local-finance-initiative/state-and-local-background ers/highway-and-road-expenditures.

225 *and we spend $80 billion a year on state jails and federal prisons:* "Economics of Incarceration," Prison Policy Initiative website, https://www.prison policy.org/research/economics_of_incarceration/.

225 *One of the most comprehensive looks at these costs:* Donal P. McCarthy et al., "Financial Costs of Meeting Global Biodiversity Conservation Targets: Current Spending and Unmet Needs," *Science*, Nov. 16, 2012.

225 *The contradiction in conservation is that the public:* Erin Waite, "Many People Want to Set Aside Half of Earth as Nature," *National Geographic*, Sept. 17, 2019; "National Poll Results: How Americans View Conservation," Pennsylvania Land Trust Association, WeConservePa.org, https:// conservationtools.org/guides/111-national-poll-results.

226 *"In the end, says Archbold Station's Swain, all conservation is local":* Interview with Hilary Swain, Lake Placid, Fla., Feb. 16, 2021.

226 *It dates back to World War II:* "The Story Behind 'Coming in on a Wing and a Prayer,'" *Famous Quotations*, Feb. 27, 2021, http://www.thisdayin quotes.com/2010/02/comin-in-on-wing-and-prayer.html.

227 *"If we can save birds":* Video interview with Pete Marra, Feb. 12, 2021.

Afterword: How We Can Help

230 *up to a billion birds die every year:* Scott R. Loss, Tom Will, Sara S. Loss, and Peter P. Marra, "Bird–Building Collisions in the United States: Estimates of Annual Mortality and Species Vulnerability," *The Condor* 116, no. 1 (February 1, 2014): 8–23, https://doi.org/10.1650/CONDOR-13-090.1.

230 *Many cities have "lights out nights":* "Lights Out: Providing Safe Passage for Nocturnal Birds," National Audubon Society website, https://www .audubon.org/lights-out-program.

230 *start a letter-writing campaign:* "Lights Out—Elected Official Sample Letter," a link provided on the "Lights Out: Providing Safe Passage for Nocturnal Birds" section of the National Audubon Society website, https://vancouveraudubon.org/2021/09/30/lights-out-providing-safe -passage-for-nocturnal-migrants/.

230 *when a cat brings you a dead animal:* Amy Shojai, "Why Do Cats Bring Their Owners 'Gifts' of Dead Animals?," Spruce Pets website, Aug. 31, 2021, https://www.thesprucepets.com/cat-hunting-gifts-553946.

230 *outdoor cats are estimated to kill more than 2.6 billion birds annually:* "7 Simple Actions to Help Birds," #BringBirdsBack website, https:// www.3billionbirds.org/7-simple-actions; "Threats to Birds," U.S. Fish and Wildlife Service website, https://www.fws.gov/library/collections/threats -birds. The agency's estimates for mortality from cats range from 1.4 to 3.9 billion. The most detailed accounting can be found in Peter P. Marra and Chris Santella, *Cat Wars: The Devastating Consequences of a Cuddly Killer* (Princeton: Princeton University Press, 2016), Chapter Four.

230 *create an outdoor "catio" enclosure:* "Safe solutions for Pet Cats," American Bird Conservancy website, https://abcbirds.org/catio-solutions-cats/.

231 *One of the biggest challenges shorebirds face:* María Paula Rubiano A., "Why Leashing Dogs Is an Easy Way to Protect Birds and Their Chicks," *Audubon Magazine*, Aug. 18, 2020, https://www.audubon.org/news/why -leashing-dogs-easy-way-protect-birds-and-their-chicks.

232 *Three quarters of the world's coffee farms:* "About Bird Friendly Coffee," National Zoo website, https://nationalzoo.si.edu/migratory-birds/about -bird-friendly-coffee.

232 *The Smithsonian Migratory Bird Center started a "Bird Friendly" certifica-tion:* "Smithsonian Bird Friendly," National Zoo website, https://nation-alzoo.si.edu/migratory-birds/bird-friendly.

233 *a small, women-owned business that donates a portion of its profits:* Bird Collective.com, https://www.birdcollective.com/pages/contribution.

234 *show them how to use the Cornell Lab's free Merlin Bird ID:* "Identify the Birds You See or Hear with Merlin Bird ID," Cornell Lab of Ornithology Merlin website, https://merlin.allaboutbirds.org/.

234 *a "Wirecutter" review team:* Kit Dillon, "The Best Bird Feeders," *New York Times*, April 29, 2022, https://www.nytimes.com/wirecutter/reviews/best-bird-feeders/.

235 *Just mix 1 part sugar to 4 parts water:* "Hummingbird Nectar Recipe," Smithsonian's Migratory Bird Center website, https://nationalzoo.si.edu/migratory-birds/hummingbird-nectar-recipe.

235 *boxes made of unpainted wood that's stained with a natural wood preservative:* "Birdhouses," Massachusetts Audubon website, https://www.massaudubon.org/learn/nature-wildlife/birds/birdhouses.

235 *three dozen types of cavity-nesting birds:* "Choose the Right Birdhouse," National Wildlife Federation website, March 2, 2011, https://www.nwf.org/Magazines/National-Wildlife/2010/Best-Bird-Houses.

236 *In 2021, Audubon facilitated more than 170,000 people:* "Top Wins for Birds and People in 2021," National Audubon Society website, Nov. 29, 2021, https://www.audubon.org/news/top-wins-birds-and-people-2021.

236 *The Cornell Lab also offers a free course on how to use eBird:* "eBird," Cornell Lab of Ornithology's eBird website, https://ebird.org/about.

236 *started its Garden for Wildlife program in 1973:* "About," National Wildlife Federation's Garden for Wildlife website, https://www.nwf.org/Garden-for-Wildlife/About.

236 *Caterpillars, for example, are a critical food source:* "About This Tool," National Wildlife Federation's Native Plant Finder website, https://www.nwf.org/nativeplantfinder/About#about.

237 *Garden for Wildlife also offers another chance:* "Shop Native Plants by State," National Wildlife Federation's Garden for Wildlife website, https://gardenforwildlife.com/pages/shop-native-plants-by-state.

237 *Leave dead trees where they lie:* Shannon Trimboli, "Leave Standing Trees (When Safe)," Backyard Ecology website, May 5, 2020, https://www.backyardecology.net/leave-standing-dead-trees-when-safe/.

237 *insecticides, the neonicotinoids or neonics:* Oliver Milman, "Fears for Bees as US Set to Extend Use of Toxic Pesticides That Paralyze Insects," *The Guardian,*

March 8, 2022, https://www.theguardian.com/environment/2022/mar/08/us-epa-toxic-pesticides-paralyse-bees-insects.

237 *There are eighty varieties of seabirds that are especially vulnerable:* Chris Wilcox, Erik Van Sebille, and Britta Denise Hardesty, "Threat of Plastic Pollution to Seabirds Is Global, Pervasive, and Increasing," PNAS website, Aug. 31, 2015, https://www.pnas.org/doi/full/10.1073/pnas.1502108112.

238 *These groups, all top rated by Charity Navigator:* Verified performances with Charity Navigator and GuideStar are as follows:

American Bird Conservancy:
https://www.charitynavigator.org/index.cfm?keyword_list=american+bird+conservancy&bay=search.results.

National Audubon Society:
https://www.charitynavigator.org/index.cfm?keyword_list=national+audubon&bay=search.results.

Bird Conservancy of the Rockies:
https://www.guidestar.org/profile/84-1079882.

Cornell Lab of Ornithology (Donations made to the lab are distributed through Cornell University via the lab's membership and donations portals on its website. The Friends of the Cornell Lab is a private nonprofit that has supported the lab but is not a part of the lab.):
https://www.charitynavigator.org/index.cfm?keyword_list=Cornell+university&bay=search.results.

241 *"I have read or been told time after time":* Daniel J. Lebbin, Michael J. Parr, and George H. Fenwick, *The American Bird Conservancy Guide to Bird Conservation* (Chicago: University of Chicago Press, 2010), iv.

Selected Bibliography

Ackerman, Jennifer. *The Bird Way: A New Look at How Birds Talk, Work, Play, Parent and Think.* New York: Penguin, 2020.

———. *The Genius of Birds.* New York: Penguin, 2016.

Beatley, Timothy. *The Bird-Friendly City: Creating Safe Urban Habitats.* Washington, D.C.: Island Press, 2020.

Cabin, Robert J. *Restoring Paradise: Rethinking and Rebuilding Nature in Hawai'i.* Honolulu: University of Hawai'i Press, 2013.

Carson, Rachel. *Silent Spring.* New York: Houghton Mifflin, 1962.

Clark, Tim W., Richard P. Reading, and Alice L. Clarke, eds. *Endangered Species Recovery: Finding the Lessons, Improving the Process.* Washington, D.C.: Island Press, 1994.

Cokinos, Christopher. *Hope Is a Thing with Feathers: A Personal Chronicle of Vanished Birds.* New York: Jeremy P. Tarcher/Putnam, 2000.

Davis, Jack E. *The Bald Eagle: The Improbable Journey of America's Bird.* New York: W. W. Norton, 2022.

———. *An Everglades Providence: Marjory Stoneman Douglas and the American Environment.* Athens: University of Georgia Press, 2009.

Franzen, Jonathan. *The End of the End of the Earth: Essays.* New York: Farrar, Straus & Giroux, 2018.

Furtman, Michael. *The Ducks Unlimited Story: Conservation for Generations.* Memphis: Ducks Unlimited, 2011.

Gallagher, Tim. *The Grail Bird.* New York: Houghton Mifflin, 2006.

Graham, Frank Jr. *The Audubon Ark: A History of the National Audubon Society.* New York: Alfred A. Knopf, 1990.

Lebbin, Daniel J., Michael Parr, and George H. Fenwick. *The American Bird*

Conservation Guide to Bird Conservation. Chicago: University of Chicago Press, 2010.

Lederer, Roger. *Beaks, Bones & Bird Songs: How the Struggle for Survival Has Shaped Birds and Their Behavior.* Portland, Ore.: Timber Press, 2016.

Leer, Linda. *Rachel Carson: Witness for Nature.* New York: Houghton Mifflin Harcourt, 1997.

Lovette, Irby J., and John W. Fitzpatrick. *Handbook of Bird Biology.* West Sussex, U.K.: Princeton University Press, 2004.

Marra, Peter P., and Chris Santella. *Cat Wars: The Devastating Consequences of a Cuddly Killer.* Princeton: Princeton University Press, 2016.

Prum, Richard O. *The Evolution of Beauty: How Darwin's Forgotten Theory of Mate Choice Shapes the Animal World—and Us.* New York: Doubleday, 2017.

Rich, Nathaniel. *Second Nature: Scenes from a World Remade.* New York: Macmillan, 2022.

Ridgely, Robert S., and Paul Greenfield. *The Birds of Ecuador.* Ithaca, N.Y.: Cornell University Press, 2001.

Rosen, Jonathan. *The Life of the Skies: Birding at the End of Nature.* New York: Farrar, Straus & Giroux, 2008.

Sekercioglu, Cagan H., Daniel G. Wenny, and Christopher J. Whelan, eds. *Why Birds Matter: Avian Ecological Function and Ecosystem Services.* Chicago: University of Chicago Press, 2016.

Shapiro, Beth. *How to Clone a Mammoth: The Science of De-extinction.* Princeton: Princeton University Press, 2015.

Walters, Mark Jerome. *Florida Scrub-Jay: Field Notes on a Vanishing Bird.* Gainesville: University Press of Florida, 2021.

Weidensaul, Scott. *A World on the Wing: The Global Odyssey of Migratory Birds.* New York: W. W. Norton, 2021.

Yaffee, Steven Lewis. *The Wisdom of the Spotted Owl: Policy Lessons for a New Century.* Washington, D.C.: Island Press, 1994.

Photo Credits

Photos Throughout Text

Courtesy of Anders Gyllenhaal—pages viii, 10, 32, 50, 62, 84, 90, 104, 122, 127, 190, 216, and 228

Graphic by Adam Smith—page 43

Courtesy of Jack Jeffrey—page 144

Courtesy of Jeremy Roberts/Conservation Media—page 170

Photos in Book Insert

Courtesy of Anders Gyllenhaal—images # 1, 2, 3, 4, 5, 6, 8, 9, 10, 11, 12, 13, 14, 15, 17, 18, 19, 20, 22, and 23

Courtesy of John Fitzpatrick—image # 7

Courtesy of Pete Cross—images # 24, 25

Index

NOTE: Page numbers beginning with 248 refer to notes. Numbers in *italic* refer to captions.

Africa, 86, 92

Agriculture Department, U.S., 48, 130, 174, 181, 225
    Working Lands for Wildlife program of, 182

'Akeke'e, 156

'Akikiki, 156, 158

Alaska, 53, 141, 149, 193

Amazon.com, 232

Amazon rainforests, 125, 137, 173, 240

American Bird Conservancy (ABC), 37, 40, 46, 48, 126, 138, 139, 148, 150, 151, 153, 214, 223, 230, 233, 235, 236, 238, 241, 295

American Crows, 106

American Eagle Foundation, 238

American Redstarts, 56

American Robins, 56

Andean Cock-of-the-Rocks, 136

Andean Condors, 128

Andes Mountains, 58, 59, 118, 125, 137

Antolini, Denise, 155

Appalachian Mountain Joint Venture, 129

Appalachian Mountains, 123, 125, 128, 129–30, 142

Aransas National Wildlife Refuge, 76

ARC (Archipelago Research and Conservation), 152

Archbold, Richard, 65

Archbold Biological Station, 12, 58, 59, 64, 65, 66, 69, 70, 71, 101, 184, 185, 224, 226

Argentina, 214, 219, 220

Arkansas, 21, 22, 23, 27

Arkansas Audubon Society, 23

Arkansas Game and Fish Commission, 23

Army, U.S., 201, 202, 224, 226

Army Rangers, U.S., 201

Ashcraft, Hugh G., Jr., 226–27

Ashe, Dan, 213, 222–23, 258

Audubon Birding Adventures for Kids (Audubon Society), 231

Audubon, John James, 72

Audubon Florida, 19, 68

Audubon Societies, 55

Austin, Tex., 39

Bald Eagles, 13, 31, 51–53, 58, 60–61
    as conservation success story, *50*, 52, 53, 60
    lead poisoning as threat to, 222, 257–58
    tracking and counting of, 52–53, 106
    wind turbines as threat to, 112

Baltimore Orioles, 102

Band-tailed Pigeons, 166, 167

Barn Owls, 151

Barred Owls, 93, 97–98, 101–103
    distinctive hoots of, 95, 101–102, 103
    as intrusive species in western states, *90*, 91, 95, 98, 100, 102–103

Barred Owls (*cont.*)
  physical appearance of, 102
  shooting and removal of, 98–100, 103
Barry, Don, 212, 222
Baryshnikov, Mikhail, 14–15
Bass Pro Shops, 200
Bayou DeView waterway, 21
Benirschke, Kurt, 161
Bernhardt, David, 209
Bezos, Jeff, 137
Biden, Joe, 95
Biden administration, 281
Biesemier, Sarah, 19–20
Bird Banding Lab (Patuxent Research
    Refuge), 72–74, 82
BirdCast, 106
Bird Collective, 233
Bird Conservancy of the Rockies, 238, 295
Bird-Friendly Beef, 232
"Bird-Friendly" certification (Smithsonian
    Migratory Bird Center), 232
BirdLife International, 40, 138
  Americas Team of, 138
  Red List of, 254
Bird Migration Explorer (tracking site),
    107
"Bird Returns" project, 111–12
Birds of Paradise, 38
Birds of Prey Foundation, 238
Birdsong Project (Audubon Society), 233
BirdSource, 108
Black-billed Magpies, 188
Blackburnian Warblers, 126
Black-necked Stilts, 186
Bloomberg, Mike, 137
Blue Jays, 64, 129
Bobolink Foundation, 179
Bogotá, Colombia, 118
Bohlman, Herman T., 120
Bolivar Peninsula, 141
Bowman, Reed, 12, 63, 64, 65–66, 68, 70–72
Bradley, Guy, 54–55
Brand, Stewart, 164–65, 167
Brennan, John, 111
Brevard County, Fla., 68
Brinkley, Ark., 23
British Army, 113
Brown Pelicans, 55
Bruneau Dunes State Park, 187–88

Buck Island Ranch, 184–85, 186, 224
Bunyan, Paul, 93
Bush, George W., 179
Bystrak, Danny, 56, 57

Cache River National Wildlife Refuge, 21
California, 9, 92, 93, 101, 149, 157, 162, 186
  Barred Owls as invasive species in, 90, 91,
    95, 98, 100, 102–103
  Barred Owls removal project in, 98–100,
    103
  "Bird Returns" project in, 111–12
  Central Valley in, 89, 100, 109–111
  Sierra Nevada mountains in, 2, 87,
    88–91, 90, 96–98, 99–100, 101
  urbanization and development in, 110
California Condors, 13, 55, 162, 225
California Rice Commission, 112
California Spotted Owls, 1–2, *84*, 87, 88,
    93, 95–97, 225
  Barred Owls as threat to, 90, 91, 98,
    99–100, 102–103
  bioacoustics in surveying population
    numbers of, 87, 88, 89–92, 96–98, 105
  declining numbers of, 89, 91, 100,
    264–265
  natural habitat of, 89, 96, 100
Canada, 36, 37, 39, 44, 46, 53, 80, *90*, 93, 95,
    128, 193, 214, 230
Canadian Wildlife Service, 33
Cape Canaveral, 15
Caribbean, 16
Carolina Chickadees, 236
Carolina Parakeets, 165
Carson, Rachel, 47–48
Cedar Waxwings, *228*, 233
Celis-Murillo, Antonio, 72–73, 74, 82
Center for Avian Population Studies
    (Cornell Lab), 108, 112, 185
Center for Biological Diversity, 147, 181,
    213, 238
Central America, 115, 138
Central Valley, Calif., 89, 100, 109–111
  "Bird Returns" project in, 111–112
  unique geology of, 110
Cerulean Warblers, *122*, 123–25, 128–33,
    137, 214
  authors' first sightings of, 140–42
  calls and songs of, 140

declining numbers of, 125, 130–31
long migratory journey of, 124, 131, 133, 141
natural habitat, 123, 128, 130
nesting by, 123–24, 142
North American conservation efforts for, 129–33, 139
Challenger (trained eagle), 52
Charity Navigator, 238, 295
Child, Penny, 23
Chocó Corridor, 133–35, 137
Chu, Miyoko, 45–46
Churchill, Winston, 195
Cincinnati Zoo, 165
Clark, Jamie Rappaport, 212, 213
Clinton, Bill, 14, 94
Cochran, Bill, 57
Collier's, 196
Colombia, 114, 118, 128, 132, 135, 138
Congress, U.S., 37, 48, 54, 55, 139, 193, 221, 223, 235–36
Conserva Aves project, 138
Conservation Birders, 241, 242
Cooper, Nathan, 64
Cornell Lab of Ornithology, 5, 22, 23, 24, 35–36, 37, 42, 45, 46, 48, 49, 59, 61, 65, 114, 115, 210, 213, 223, 231, 233, 238, 241, 295
Center for Avian Population Studies at, 108, 112, 186
eBird project of, see eBird
efforts to track pollinators at, 116
July, 2017, gathering at, 35, 36, 39–40, 41–42, 44
K. Lisa Yang Center for Conservation Bioacoustics at, 36, 86
Macaulay Library at, 36
Merlin Bird ID smartphone app of, 234
Project FeederWatch of, 236
Costa Rica, 114, 128
Crampton, Lisa "Cali", 157–58, 159, 219
Crespo, Juan Carlos, 126, 128
Crimson-rumped Toucanets, *127*
CRISPR (gene-editing tool), 166
Cruz, George, 124–25
Culex mosquito, 145–46, 157
Currituck Inlet, 195
Currituck Sound, 194, 195, 197

Darwin, Charles, 125
Davis Ranches, 111, 186
"Dawn of De-Extinction, The: Are You Ready?" (Brand TED Talk), 164, 165
Dayer, Ashley, 200, 211
Defenders of Wildlife, 212
Defense Department, U.S., 48
conservation efforts of, 201–2, 203–4, 209, 224
Delaney, Michael, 16
Democratic Party, 95, 223
DePriest, Gene, 23
DiCaprio, Leonardo, 137, 139
Dickcissels, 186, 239
Dickinson, Emily, 217
Disney World, 10, 12, 16
Doane, Michael, 116–117, 186
Dokter, Adriaan, 41, 42, 43, 106
Donner, Amy, 107
Doresky, John, 3
Droll Yankee Onyx Mixed Seed Tube Bird Feeder, 234
Duck Stamp, 199–200, 248
Ducks Unlimited (DU), 48, 191–92, 193, 194, 196–97, 198–99, 200, 215, 223, 238
founding of, 194, 196
fundraising banquets and tournaments of, 191–92, 193, 194
Dugway Proving Ground, 202
Dunlins, 186
Dupont family, 195
Dusky Seaside Sparrows, 15–16
Dust Bowl, 55, 192

Earth Fund, 137
Eastern Bluebirds, *viii*
Eastern Ecological Science Center (Patuxent Research Refuge), 72
Eastern Meadowlarks, 34
calls and songs of, 32, 174
declining numbers of, 32, 174
Eastern Phoebes, 6–7, 72
Eastern Towhees, 73
Eastern Wood-Pewees, 6–7
eBird, 52–53, *104*, 105–10, 112–13, 114, 119, 236
competitive appeal of, 109, 117, 118
development of, 108–9

eBird (*cont.*)
   in guiding conservation efforts, 109–10, 111–12, 115–16, 118, 120
Ecom Agroindustrial Corporation, 115
Ecuador, 124–27, 132
   biodiversity of, 124–26, 128, 133
   bird tourism in, 135–37
   conservation efforts in, 133–37, 138, 270
   large and diverse bird population of, 124–25, 126, 127, 128, 133
   Mindo region of, 126
Edison, Thomas, 85
Eisenhower, Dwight, 195
Eldermire, Charles, 61
Elvis (canary), 113
Endangered Species Act (ESA; 1973), 13, 23, 31, 48, 52, 68, 91, 93, 125, 149, 151, 202, 203, 221, 291
   debate over status of Sage Grouses under, 180, 281
   funding issues of, 147, 212, 213, 226
   lack of reauthorization of, 223
   Northern Spotted Owl controversy and, 93–94, 95
   petition to protect Ceruleans under, 130–132
   vague listing criterion of, 131
   Whooping Crane protection under, 78–79
Endangered Species List, 29, 67, 131, 209, 253, 275
   lengthy timeframe for inclusion on, 13, 94, 132, 223
Environmental Protection Agency, 39, 48
Eskimo Curlews, 165
Eurasian Magpies, 70
Evening Grosbeaks, 214
Everglades, 185

Fabritius, Carel, 217
Facebook, 86, 178
Farm Bill, 130, 181
Farmer, Chris, 148, 153–54, 156
Fearer, Todd, 129, 130, 133
Fenwick, George, 241
Finley, William L., 120
Fish and Wildlife Service, U.S., 3, 7, 16, 27, 41, 47, 52, 76, 93, 94, 112, 125, 146, 147, 154, 208, 209, 210, 215, 253, 258, 275, 281, 293
   backlog of past-due ESA rulings at, 132
   Barred Owl removal project of, 98–100
   cloning projects of, 162
   Duck Stamp and, 199–200, 248
   Fort Benning wildfire investigated by, 207, 208
   Fort Bragg shutdown by, 203
   funding issues of, 48, 132, 212–13
   ineffectual conservation policies of, 212–13, 222–23, 291
   and petition for federal protection of Ceruleans, 130, 131
   in search for Ivory-billed Woodpecker, 22, 23, 24, 29–30
Fitzpatrick, John, 5, 22, 24, 30, 58–61, 64, 65, 69, 71, 213
   eBird project and, 108, 109
   on sanctity of birds, 219–20
Flint Hills, Kans., 171, 172, 175, 176, 179
Florida, 9, 11, 15, 16, 24, 51, 54, 58–59, 76, 184
   ambivalence toward Scrub-Jay conservation in, 68–70
   development and loss of habitats in, 59, 65, 67–69, 71
   Everglades in, 185
   prairie grasslands of, 1, 12, 15, 17–19, 20, 119, 212
   scrublands of, 63, 64, 65, 66, 67–68, 69–70, 71
Florida Audubon Society, 54
Florida Bay, 54–55
Florida Cattlemen's Association, 185
Florida Dusky Seaside Sparrow, 146
Florida Fish & Wildlife Research Institute, 17
Florida Grasshopper Sparrows, 1, 2, 11–15, 16–20, 21, 119, 184, 212, 225
   calls and songs of, 12, 19–20
   captive breeding program for, 15, 16–20
   declining numbers and threat of extinction to, *10*, 12–13
   prairie grassland habitat of, 12, 17–18, 19, 20, 119
Florida International University, 16
Florida Scrub-Jays, 39, 58, 59, 63–72, 184, 225
   declining numbers of, 65, 71

development and urbanization as threat
  to, 65, 67–69, 71
fire's role in lives of, 71
natural habitat of, 63, 65, 67–68, 69–70,
  71
optimistic outlook for, 71–72
socialization and mating habits of, 70–71
tracking of, 62, 64–66, 70, 71
Florida Wildlife Corridor Act (2021), 70
Flying H ranch, 171
Flying Lessons.US (website), 7, 229, 230
  Birding Basics section of, 233
Flying Tigers (film), 227
Flying W ranch, 176, 177–79, 186–87
Ford, Bob, 131
Ford, Jerome, 112, 212
Foreign Service, U.S., 118
Forest Service, U.S., 86, 88, 101, 206
Forsman, Eric, 93, 95
Forsyth, Adrian, 241
Forsyth, Sharon Pitcairn, 238–39, 240–41,
  242
Fort Benning, 201, 202–3, 204–5, 206–9,
  226
  wildfire at, 207–8
Fort Bragg, 203
Fort Hood, 202
Frazer, Gary, 291
Freudenthal, David, 200
Frozen Zoo (San Diego Wildlife Alliance),
  160–63

Galápagos Islands, 125–26, 137
Galveston Island State Park, 142
Game of Thrones series, 100
Garden for Wildlife program (National
  Wildlife Federation), 236–37
Geological Survey, U.S., 56, 72, 153, 157,
  213, 258
  Bird Banding Lab of, 72–74, 82
Georgetown University, 44, 72, 210
Georgia, 9, 11, 201
Germany, 81, 85, 226
Giant Antpittas, 135
Gill, Frank, 108
Gilman, Howard, 14–15
Gilman Paper Company, 14
Golden-cheeked Warblers, 38–39, 202
Golden Eagles, 202, 222, 257–58

Golden-winged Warblers, 214
Goldfinch, The (Tartt), 217
Gold Rush, 93
Golet, Greg, 112
Google, 86
Gordon and Betty Moore Foundation, 137
Gore, Al, 14, 94
Grammy Hall of Fame Award, 217
Grassland Bird Trust, 233
Gray, Elizabeth, 49, 220, 241
Gray, Paul, 19
Great Antpittas, 136
Great Blue Herons, 60, *104*, 119, 120
Great Britain, 113
Greater Sage-Grouses, 139, *170*, 183
  conservation efforts for, 180–84
  declining numbers of, 174, 182
  mating dance of, 170, 174, 180, 183
Great Plains, 174
Greenfield, Paul, 124, 136, 137, 138
Greenwald, Noah, 147, 181
Griffiths, Tim, 182
Guggenheim Partners, 15
GuideStar, 295
Guide to Bird Conservation (American Bird
  Conservancy), 241

Haikubox, 107
Hakalau Forest National Wildlife Refuge,
  153, 154–55, 156
Haldane, John Scott, 113
Hawai'i, 9, 29, 53, 86, 92, *144*, 145–60,
  218–19, 291
  bleak conservation outlook for, 147
  devastated ecosystem in, 146, 147, 148,
    152–53, 154
  mass and rapid extinction of birds in,
    146, 147, 154, 156, 253, 274, 275
  mosquito project in, 3, 145–46, 148,
    156–60
  Palila conservation efforts in, 153–54,
    156
  powerlines in, 149, 150–51
Hawai'i (Big Island), 148, 152–56, 202
  Mauna Kea on, see Mauna Kea
Hawai'i, University of, 155
Hawaiian Petrels, 149–51
HawkWatch International, 233
Henslow's Sparrows, 179

High Island, Tex., 141
Smith Oaks Sanctuary on, 141–42
Hite, Justin, 146, 158
Hofstadter, Danny, 98–100
Hooded Grebes, 219–20
Hoover, Herbert, 195
" 'Hope' Is the Thing with Feathers"
(Dickinson), 217
Horned Larks, 34
Houck, Marlys, 161–62
House of Representatives, U.S., 200
House Sparrows, 8, 106
Houston Audubon Society, Smith Oaks
Sanctuary of, 141–42
*How to Clone a Mammoth: The Science of
De-Extinction* (Shapiro), 161
Hoy, Gwen, 172, 173, 174, 175, 177,
178–79, 186
background of, 175–76
Hoy, Josh, 174, 175, 176–80, 186–87, 189
background of, 171–73
Hoy, Josie, 175, 177, 178–80
Hugo, Hurricane, 206

India, 112, 219
Interior Department, U.S., 22, 48, 57, 116,
291
International Cooperation for Animal
Research Using Space (ICARUS),
80–81, 82
International Crane Foundation, 238
International Space Station, 53, 64, 80
Ivory-billed Woodpeckers, 13–14, 21–30,
165
calls and songs of, 21, 24
excitement over rediscovery of, 22–23
natural habitat of, 22, 26
physical appearance of, 21, 27, 28
rare sightings of, 21, 22, 25, 27–28, 29, 30
recordings of, 24, 29
searches for proof of existence of, 22, 23,
24–29, 30
'I'iwi honeycreepers, *144*, 156

Jackson, Mahalia, 217
Jay Watch (fan club), 68
Jefferson Memorial (Washington, D.C.), 52
Jefferson National Forest, 129
Jeffrey, Jack, 154–55

Jensen, Brian, 182
Jocotoco Conservation Foundation, 126,
133, 134, 135
Johanns, Mike, 22
Johnson, Marshall, 173, 179
Johnson, Will, 191–92, 199
Jurassic Park (film), 165

Kaestner, Peter, 117–18
Kansas, 211
tallgrass prairie of, 176–78, 179, 186–87,
189
Kansas Aldo Leopold Conservation Award,
178
Katzner, Erin, 219
Katzner, Todd, 258
Kaua'i, 145, 146, 148, 149, 150–52, 157,
218–19
powerlines on, 149, 150–51
Kaua'i Endangered Seabird Recovery
Project, 149
Kaua'i Forest Bird Recovery Project, 157,
219
Kaua'i Island Utility Cooperative, 150–51,
276
Kauka, Sabra, 159
Keitt, Brad, 150, 151
Kelly, Kevin, 96, 97–98
Kerr Scott Building (Raleigh, N.C.), 191
Kildeers, 186
Kitzes, Justin, 86–87
Klinck, Holger, 86
K. Lisa Yang Center for Conservation
Bioacoustics (Cornell Lab), 36, 86
Knapp, Joseph Palmer, 194, 195–96
Knobloch Family Foundation, 215
Knotts Island, N.C., 195–97
Kryshak, Nick, 98–100
Kurt (cloned Przewalski's horse), 162

Lacey Act (1900), 54
Lake Okeechobee, 185
Lake Wales Ridge, 69
Lamberski, Nadine, 4, 162–63
LaPointe, Dennis, 156–57, 160
Latin America, 16, 80, 124, 126, 128, 129,
139, 223
efforts to fund and protect land in, 126,
138, 270

see also South America; specific
   countries
Latta, Steve, 25, 26–27, 29, 30
Lebbin, Dan, 126–28, 133
Leonardo da Vinci, 220
Leopold, Aldo, 176
Lesser Yellowlegs, 214
Light-footed Ridgeway's Rail, 202
"Lights Out for Birds" nights, 230
Liverpool, England, 40
Los Angeles Dodgers, 15
Lotus (Bald Eagle), 61
Louisiana, 13, 77, 188
   secret search for Ivory-billed
      Woodpecker in, 24–29, 30
   wetlands in, 76–77, 79
   Whooping Crane population in, 75, 76,
      78, 79
Louisiana Department of Wildlife and
   Fisheries, 75
Louisiana Waterthrush, 105

Macey, John, 202
Mackay Island National Wildlife Refuge,
   196–97
Mackenzie, Stuart, 82
Madison, Brian, 194
Malta, 148
Māmane trees, 153, 154
Mann, David, 107
Marra, Pete, 43–45, 48, 58, 72, 198, 227
   on need to rethink conservation, 210,
      211, 212, 215
Marsden, Tim, 203
Martha (last Passenger Pigeon), 165
Maui, Hawai'i, 145
Mauna Kea, 148, 152–56
   Hakalau Preserve on, 153, 154–55, 156
   Palila conservation efforts on, 153–54,
      156
Merlin Bird ID (smartphone app), 234
Metropolitan Life Insurance Company, 196
Mexican Spotted Owls, 93, 264–65
Mexico, 93, 124, 141, 162, 193
Mexico, Gulf of, 124, 133, 141
Michaels, Mark, 25, 27, 30
Michot, Tommy, 27, 28
Migratory Bird Treaty Act (1918), 55, 221
Mississippi, 119, 187

Mississippi River, 21, 76, 141, 173
Mona Lisa (Leonardo da Vinci), 219, 220
Monsanto, 117
Montana, University of, 31, 181, 214
More Game Birds in America Foundation,
   196
Morris, John, 200
Motus Wildlife Tracking System, 80–82
Mr. President (Bald Eagle), 61
Murdock, Madeleine, 182–184

Nanegalito, Ecuador, 136
National Aeronautics and Space
   Administration (NASA), 101, 114,
   187
   "Soundscapes to Landscapes" project of,
      114
National Arboretum, U.S., 60–61
National Audubon Society, 39, 48, 49, 59,
   93, 138, 173, 179, 194, 212, 213, 220,
   223, 231, 236, 238, 241, 295
   Bird-Friendly Beef brand of, 232
   Birdsong Project of, 233
   eBird project and, 106–10
   online Marketplace of, 233
   state and local chapters of, 19, 54, 55,
      68, 235
   and upcoming conservation legislation,
      236
National Aviary, 24, 25, 113
National Geographic, 120
National Guard, U.S., 146
National Public Radio, 23
National Science Foundation, 107, 108, 213
National Wildlife Federation, 238
   Garden for Wildlife program of, 236–37
National Wildlife Refuge Association, 238
National Wildlife Refuges, 55, 197, 198,
   233
   see also specific wildlife refuges
Native Plant Finder (website), 236
Nature Conservancy, The, 23, 110–111,
   116–117, 186, 194, 238
   "Bird Returns" project of, 111–112
Naugle, Dave, 31, 181, 214
NBC News, 46
Nene geese, 86, 202
Neotropical Migratory Bird Conservation
   Act (2002), 139

Nespresso, 114, 115, 224
Newell's Shearwater, 149–51
New Yorker, 47
New York Times, 46, 47, 234
Nicaragua, 115
North American Bluebird Society, *viii*, 238
North American Breeding Bird Survey,
    56–57
North American Waterfowl Management
    Plan, 197
North American Wetlands Conservation
    Act (1989), 197, 221
North Carolina, 147, 191, 194, 197, 203,
    206, 240
North Carolina State University, 205
Northern Spotted Owls, 87, 92–95, 98,
    264–65
    declining numbers of, 95
    "timber wars" ignited over, 87, 93, 94–95
Northwest Forest Plan, 94–95
Norton, Gail, 22
Novak, Ben, 163–64, 166–68

Officer Candidate School, 201
Oliver-crowned Yellowthroats, 126
Oregon, 93, 94, 95
Orndorff, Ryan, 204, 224
Ospreys, 31, 55
Oteyza, Juan, 17–18, 20
Owens, Ian, 49, 112, 116, 241
Oyster Key, Fla., 54–55

Pakistan, 219
Palila honeycreepers, 153–54, 156
Panama, 80, 128
Panama Canal, 80
Parker, James, 202, 203, 204, 206, 207–9
Parr, Mike, 37, 39–41, 43, 46–47, 126–28,
    139, 214
Partners in Flight, 36–37
Partners in Flight Science Committee,
    36–37
Passenger Pigeons, 5, 147, 164–65
    ecological benefits of, 167–68
    rapid decline and extinction of, 165
    rebuilding project for, 163–165, 166–169
Patuxent Research Refuge, 56
    Bird Banding Lab at, 72–74, 82
    Eastern Ecological Science Center at, 72

Paulson, Wendy and Hank, 179
Paz, Angel, 135, 136, 137
Paz, María, 136
Peery, Zach, 88, 91, 99, 101
Penniman, Teya, 159
Pepper, Justin, 179
Peregrine Falcons, 31, 55
Peregrine Fund, 219
Persian Gulf War, 113
Peru, 58, 128
Phelan, Ryan, 163, 164, 169
Photo Ark project, 120
Pileated Woodpeckers, 21, 27, 28, 29
Pinedale, Wyo., 183
Pittman-Robertson Wildlife Restoration
    Act (1937), 193
Pittsburgh, University of, 24, 27, 28, 86
Point Blue Conservation Science, 111, 114
Poncon, Eric, 115
Po'ouli, 154
Portland, Oreg., 94
Powdermill Nature Reserve, 86
Project FeederWatch, 236
Project Principalis, 25–29, 30
Przewalski's horse, 162
Puaiohi thrushes, 218–19
Putnam, Adam, 193–94, 198–99

Quito, Ecuador, 128

Raine, André, 148–49, 150, 151–52
Raine, Helen, 148–49, 151–52
Raleigh, N.C., 140, 191
Ramos, Aurelio, 138
Rare Species Conservatory Foundation,
    16–17
Re:wild, 137, 139
Red-cockaded Woodpeckers, 39, 201,
    202–3, 204–9
    Fort Benning wildfire and, 207–208
    rescue efforts for, 205–7
    resurgence of, 190, 208–209
RedLAC, 138
Red-shouldered Hawks, 18, 28
Refugio Paz de las Aves, 136
Reillo, Paul, 16, 17
Republican Party, 95, 223
Restoring America's Wildlife Act (proposed
    bill), 200

Revive & Restore, 162, 163, 164, 165, 166, 168
Reynolds, Mark, 110–11
Ringling Bros. and Barnum & Bailey Circus, 11
Road to Recovery (R2R), 192, 210–12, 213–15
  fundraising by, 214, 215
Rob and Melani Walton Foundation, 137
Robbins, Chan, 56–57
Rock Creek Park, 105, 106, 239
Rocky Mountains, 173
Rodewald, Amanda, 108, 116, 186
Roosevelt, Teddy, 120, 198
Roscosmos (Russian space agency), 81
Rose-breasted Grosbeaks, 141
Rose Festival Parade, 94
Rosenberg, Ken, 6, 37, 43, 44–45, 47, 48, 210, 222
Roundup, 237
Ruby-crowned Kinglets, 74
Ruby-throated Hummingbirds, *216*
Ruiz-Gutierrez, Viviana, 112, 115

Sacramento River, 186
Saenz, Dan, 85–86
Sage Grouse Initiative, 181–83, 225, 281
St. Marys River, 14
Salas, Leo, 114
San Diego Zoo Wildlife Alliance, 4, 147, 160, 162
  Frozen Zoo at, 160–63
San Francisco, Calif., 45
Sapsucker Woods (Ithaca, N.Y.), 35
Sartore, Joel, 120
Sauer, John, 56–57
Saving America's Pollinators Act (proposed bill), 236
Sawyer, Sarah, 88, 91–93, 96–98, 100–101, 102
Scalia, Matthew, 208
Scarlett, Lynn, 116, 139
Scarlet Tanagers, 141
Schaefer, Martin, 133–35
Schmidt, Paul, 147, 193, 199, 215
Schumann, Andrew, 16
Science, 6, 43–45, 46, 49, 58, 210, 212
Scorsese, Martin, 14

Seal Beach, 202
Second Continental Congress, 51
Senate, U.S., 200
Shapiro, Beth, 161, 165–66
"Shortchanged" (Center for Biological Diversity report), 213
Shrum, Peggy, 27–28
Shurter, Steve, 13
Sierra Nevada mountain range, 2, *90*, 96, 99–100, 101
  Barred Owls removal project in, 98–100, 103
  bioacoustics in surveying Spotted Owls in, 87, 89–92, 96–98
Silent Spring (Carson), 47–48
Sillett, Scott, 31, 115
Skip (Josh Hoy's cousin), 175
Slate, 8
Smith, Adam, 33–34, 35, 42, *43*, 57
Smith Oaks Sanctuary, 141–42
Smithsonian Migratory Bird Center, 31, 44, 64, 115
  "Bird-Friendly" certification of, 232
Sonoma County, Calif., 114
"Soundscapes to Landscapes" project (NASA), 114
South America, 9, 37, 58, 125, 126, *127*, 128, 132, 141, 142, 148
  Chocó Corridor in, 133–35, 137
  conservation efforts in, 122, 124, 133–39
  Nespresso study gauging health of coffee farms in, 114–115
  see also Latin America; specific countries
South Carolina, 9, 206
Sparling, Gene, 21, 22
Spotted Owls, 1–2
Stafford, Scott, 105, 106–7, 117
Stanislaus National Forest, 96
"State of the Birds" report (2008), 57
Steiner, Cynthia, 162
Suh, Young Ha, 66, 70
Swain, Hilary, 101–102, 184, 185, 224, 226
Swift, Byron, 137, 139

Tanner, James, 26
Tartt, Donna, 217

Texas, 39, 157, 205
This Week (newspaper supplement), 196
Three Billion Bird report, 6, 44–45, 48, 49,
    57, 58, 139, 174, 192, 193, 198, 210,
    212, 222, 227
    publicity surrounding release of, 46–47
3Billionbirds.org (website), 46
Time, 94
Trahan, Amy, 29
Trump administration, 95, 281
Tursi, Frank, 195

Utah, Dugway Proving Ground in, 202

Venezuela, 128, 239
Virginia, 6, 129, 194, 195, 205, 239
Virginia Tech, 200, 211
    Department of Biological Sciences at, 206

Walker, Norman, 113
Walmart, 116, 224
Walmart Foundation, 116
Walter, Kimbra, 15
Walter, Mark, 15
Walter Conservation, 15
Walters, Jeff, 205–6
Washington Post, 60
Washington State, 93, 95, 106
Wayne, John, 227
Western Kingbirds, 187
Western Meadowlarks, 34
Whiskered Wrens, 126
Whitehouse, Bennett, 192
White Lake Wetlands Conservation Area, 76
White Oak Conservation Center, 11–15
    captive breeding project at, 15–20
White Oak Dance Project, 15
White-throated Sparrows, 36, 106
Whole Foods, 232

Whooping Cranes, 13, 55, 225
    captive breeding of, 76
    federal protection of, 78–79
    natural habitats of, 75–79
    reintroduction project for, 78
    tracking of, 75, 76–77, 78
Whooping Crane Studbook, The (registry), 76
Wikelski, Martin, 80–81
Wild Birds Unlimited, 235
Will, Tom, 41, 210
Williams, Martha, 291
Williams, Richard, 195–97
Wilsey, Chad, 107
Wilson, Woodrow, 85
Wisconsin Historical Society, 93
Wolbachia mosquito project, 145–46, 148,
    156–60
Wood, Chris, 109
Wood, Connor, 87–91, 97–98, 99, 100, 101,
    105
Wood Thrushes, 34
Working Lands for Wildlife program
    (Agriculture Department), 182
World Bank, 270
World War I, 85
World War II, 85, 226
Worm-eating Warblers, 240
Wright brothers, 217–18

Yellow-billed Cuckoos, 214
Yellowstone National Park, 52, 182
Yellow Warblers, 56, 230
Yépez, Itala, 138
Yosemite National Park, 89, 95–96, 102
Yucatán Peninsula, 124, 141

Zambia, 148
Zimorski, Sara, 75, 76–78, 79
Zinke, Ryan, 258

About the Authors

ANDERS AND BEVERLY GYLLENHAAL are lifelong journalists who took up birding as they concluded careers as writers and editors—and soon began a new chapter researching, photographing, and following birds. Beverly started as a reporter for *The News & Observer* in Raleigh, North Carolina, followed by years as a reporter and editor at the *Miami Herald*. She then developed syndicated columns on child-rearing and cooking, which led to three Desperation Dinners cookbooks and a website. Anders held a series of reporting jobs before becoming executive editor at newspapers in Raleigh, Minneapolis, and Miami and later vice president of news for the McClatchy Company. They've since put their journalistic skills to work on behalf of birds, launching their website, FlyingLessons.US: What We're Learning from the Birds. Their work on endangered species, the evolution of avian science, and innovations in conservation has appeared in newspapers and magazines around the country. Beverly and Anders live in Raleigh, where their children, Sam and Grey, grew up. They continue to spend much of their time following the birds.